W9-AUG-215

THE ART OF
SYSTEMS
ARCHITECTING

systems engineering series

series editor
A. Terry Bahill, University of Arizona

THE ART OF SYSTEMS ARCHITECTING

Eberhardt Rechtin
University of Southern California
Los Angeles, California

Mark W. Maier
University of Alabama in Huntsville
Huntsville, Alabama

CRC Press
Boca Raton   New York   London   Tokyo

Acquiring Editor:	Norm Stanton
Project Editor:	Mary Kugler
Marketing Manager:	Susie Carlisle
Cover design:	Denise Craig

Library of Congress Cataloging-in-Publication Data

Rechtin, Eberhardt.
    The art of systems architecting / Eberhardt Rechtin and Mark Maier.
        p. cm.
    Includes bibliographical references and index.
    ISBN 0-8493-7836-2 (alk. paper)
    1. Systems engineering. I. Maier, Mark. II. Title.
TA168.R368 1997
620'.001'171—dc20

                        96-31918
                          CIP

No claim to original U.S. Government works
International Standard Book Number 0-8493-7836-2
Library of Congress Card Number 96-31918
Printed in the United States of America 1 2 3 4 5 6 7 8 9 0
Printed on acid-free paper

Preface

The continuing development of systems architecting

*Architecting, the planning and building of structures, is as
old as human societies — and as modern as the exploration
of the solar system.*

So began this book's 1991 predecessor.* The earlier work was based on
the premise that architectural methods, similar to those formulated
centuries before in civil works, were being used, albeit unknowingly,
to create and build complex aerospace, electronic, software, command,
control, and manufacturing systems. If so, then still other civil works
architectural tools and ideas — such as qualitative reasoning and the
relationships between client, architect, and builder — should be found
even more valuable in today's more recent engineering fields. Five
years later, judging from several hundred retrospective studies at the
University of Southern California of dozens of post-World War II systems, the original premise has been validated. Today's systems architecting is indeed driven by, and serves, much the same purposes as
civil architecture — to create and build systems too complex to be
treated by engineering analysis alone.

Of great importance for the future, the new fields have been creating architectural concepts and tools of their own and at an accelerating rate. This book includes a number of the more broadly applicable
ones, among them heuristic tools, progressive design, intersecting waterfalls, feedback architectures, spiral-to-circle software acquisition,
technological innovation, and the rules of the political process as they
affect system design.

Arguably, these developments could, even should, have occurred
sooner in this modern world of systems. Why now?

* Rechtin, E., *Systems Architecting, Creating & Building Complex Systems*, Prentice-Hall,
Englewood Cliffs, NJ, 1991.

Architecting in the systems world

A strong motivation for expanding the architecting process into new fields has been the retrospective observation that success or failure of today's widely publicized systems often seems pre-ordained, that is, traceable to their beginnings. It is not a new realization. It was just as apparent to the ancient Egyptians, Greeks, and Romans who originated classical architecting in response to it. The difference between their times and now is in the extraordinary complexity and technological capability of what could then and now be built.

Today's architecting must handle systems of types unknown until very recently; for example, systems which are very high quality, real time, closed loop, reconfigurable, interactive, software intensive, and for all practical purposes, autonomous. New domains like personal computers, intersatellite networks, health services, and joint service command and control are calling for new architectures — and for architects specializing in those domains. Their needs and lessons learned are in turn leading to new architecting concepts and tools and to the acknowledgment of a new formalism — and evolving profession — called systems architecting, a combination of the principles and concepts of both systems and of architecting.

The reasons behind the general acknowledgment of architecting in the new systems world are traceable to that remarkable period immediately after end of the cold war in the mid-1980s. Abruptly, by historical standards, a 50-year period of continuity ended. During the same period, there was a dramatic upsurge in the use of smart, real-time systems, both civilian and military, that required much more than straightforward refinements of established system forms. Long-range management strategies and design rules, based on years of continuity, came under challenge. It is now apparent that the new era of global transportation, global communications, global competition, and global turmoil is not only different in type and direction, it is unique technologically and politically. It is a time or restructuring and invention, of architecting new products and processes, and of new ways of thinking about how systems are created and built.

Long-standing assumptions and methods are under challenge. For example, for many engineers, architectures were a given; automobiles, airplanes, and even spacecraft had the same architectural forms for decades. What need was there for architecting? Global competition soon provided an answer. Architecturally different systems were capturing markets. Consumer product lines and defense systems are well-reported examples. Other questions: how can software architectures be created that evolve as fast as their supporting technologies? How deeply should a systems architect go into the details of all the system's subsystems? What is the distinction between architecting and engineering?

Distinguishing between architecting and engineering*

Because it is the most asked by engineers in the new fields, the first issue to address is the distinction between architecting and engineering in general, that is, regardless of engineering discipline. Although civil engineers and civil architects, even after centuries of debate, have not answered that question in the abstract, they have in practice. *Generally speaking*, engineering deals almost entirely with measurables using analytic tools derived from mathematics and the hard sciences; that is, engineering is a *de*ductive process. Architecting deals largely with unmeasurables using nonquantitative tools and guidelines based on practical lessons learned, that is, architecting is an *in*ductive process. At a more detailed level, engineering is concerned with quantifiable costs, architecting with qualitative worth. Engineering aims for technical optimization, architecting for client satisfaction. Engineering is more of a science, architecting more of an art.

In brief, the practical distinction between engineering and architecting is in the problems faced and the tools used to tackle them. This same distinction appears to apply whether the branch involved is civil, mechanical, chemical, electrical, electronic, aerospace, software, or systems. Both architecting and engineering can be found in every one.

Systems architecting is the subject of this book, and the art of it in particular, because, being the most interdisciplinary, its tools can be most easily applied in the other branches. From the authors' experience, systems architecting quickly and naturally abstracts and generalizes lessons learned elsewhere, not only for itself but for transfer and specialization in still other branches. A good example is the system guideline (or heuristic), abstracted from examples in all branches, of **Simplify. Simplify. Simplify.** It will appear several times in this text.

It is important in understanding the subject of this book to clarify certain expressions. The word "architecture" in the context of civil works can mean either a structure, a process, or a profession; in this text it refers only to the structure. The word "architecting" refers only to the process. Architect*ing* is an invented word to describe how architectures are created, much as engineer*ing* described how "engines" and other artifacts are created. In another, more subtle, distinction from conventional usage, an "architect" is meant here to be an individual engaged in the process of architecting, regardless of domain, job title, or employer. By definition and practice both, from time to time an architect may perform engineering and an engineer may perform architecting — whatever it takes to get the job done.

---

* Because the fields of systems engineering and architecting are relatively new, the definitions of many terms, including "architecting" itself, are not yet standardized. For the purposes of this book, therefore, a Glossary at the end of this book provides a list of the most important ones and their meanings as used by the authors.

Clearly, both processes can and do involve elements of the other. Architecting can and does require top-level quantitative analysis to determine feasibility and quantitative measures to certify readiness for use. Engineering can and occasionally does require the creation of architecturally different alternatives to resolve otherwise intractable design problems. For complex systems, both processes are essential. In practice, it is rarely necessary to draw a sharp line between them.

Criteria for mature and effective systems architecting

An increasingly important need of project managers and clients is for criteria to judge the maturity and effectiveness of systems architecting in their projects — criteria analogous to those developed for software development by Carnegie Mellon's Software Engineering Institute. Based upon experience to date, criteria for systems architecting appear to be, in rough order of attainment:

- A recognition by clients and others of the need to architect complex systems
- An accepted discipline to perform that function; in particular, the existence of architectural methods, standards, and organizations
- A recognized separation of value judgments and technical decisions between client, architect, and builder
- A recognition that architecture is an art as well as a science; in particular, the development and use of nonanalytic as well as analytic techniques
- The effective utilization of an educated professional cadre; that is, of masters-level, if not doctorate-level, individuals and teams engaged in the process of systems-level architecting

By those criteria, systems architecting is in its adolescence — a time of challenge, opportunity, and controversy. History and the needs of global competition would seem to indicate adulthood is close at hand.

The architecture of this book

The first priority of this book has been to restate and extend into the future the classical architecting paradigm introduced in the author's (Rechtin) earlier book. An essential part of both the classical and extended paradigms is the recognition that systems architecting is part art and part science. Part One of this book further develops the art and extends the central role of heuristics. Part Two introduces four important domains which both need and contribute to the understanding of that art. Part Three helps bridge the space between the science and the art of architecting. In particular, it develops the core architecting process of modeling and representation by and for software. Part Four concentrates on architecting as a profession: the political process and its part in system design, and the professionalization of the field through education, research, and peer-reviewed journals.

The architecture of Part Two deserves an explanation. Without one, the reader may inadvertently skip some of the domains — builder-architected systems, manufacturing systems, social systems, and software systems — because they are outside the reader's immediate field of interest. These chapters, instead, recognize the diverse origins of heuristics, illustrating and exploiting them. Heuristics often first surface in a specialized domain where they address an especially prominent problem. Then, by abstraction or analogy, they are carried over to others and become generic. Such is certainly the case in the selected domains. In these chapters the usual format of stating an heuristic and then illustrating it in several domains is reversed. Instead it is stated, but in generic terms, in the domain where it is most apparent. Readers are encouraged to scan all the chapters of Part Two. The chapters may even suggest domains, other than the reader's where the reader's experience can be valuable in these times of vocational change. References are provided for further exploration. For professionals already in one of the domains, the description of each is from an architectural perspective, looking for those essentials that yield generic heuristics and providing in return other generic ones that might help better understand those essentials. In any case, the chapters most emphatically are not intended to advise specialists about their specialties!

Architecting is inherently a multidimensional subject, difficult to describe in the linear, word-follows-word format of a book. Consequently, it is occasionally necessary to repeat the same concept in several places, internally and between books. A good example is the concept of systems. Architecting can also be organized around several different themes or threads. The author's (Rechtin) earlier work was organized around the well-known waterfall model of system procurement. As such, its applicability to software development was limited. This book, more general, is by fundamentals, tools, tasks, domains, models, and vocation. Readers are encouraged to choose their own personal theme as they go along. It will help tie systems architecting to their own needs.

Exercises are interspersed in the text, designed for self-test of understanding and critiquing the material just presented. If the reader disagrees, then the disagreement should be countered with examples and lessons learned — the basic way that mathematically unprovable statements are accepted or denied. Most of the exercises are thought problems, with no correct answers. Read them and if the response is intuitively obvious, charge straight ahead. Otherwise, pause and reflect a bit. A useful insight may have been missed. Other exercises are intended to provide opportunities for long-term study and further exploration of the subject. That is, they are roughly the equivalent of a master's thesis.

Notes and references are organized by chapter, Heuristics by tradition are bold faced when they appear alone, with an appended list of them completing the test.

A glossary is included at the end of this book to clarify the meaning of key terms as they are used in the text.

The intended readership for this book

This book is written for present and future systems architects, for experienced engineers interested in expanding their expertise beyond a single field, and for thoughtful individuals concerned with creating, building, or using complex systems.

From experience with its predecessor, the book can be used as a referenced work for graduate studies, for senior capstone courses in engineering and architecture, for executive training programs, and for the further education of consultants, systems acquisition and integration specialists, and as background for legislative staffs. It is a basic text for a Master of Science degree in Systems Architecture and Engineering at the University of Southern California.

Acknowledgments

Dr. Brenda Forman, then of USC, now of the Lockheed Martin Corporation, and the author of Chapter 10, accepted the challenge of creating a unique course on the facts of life in the national political process and how they affect — indeed often determine — architecting and engineering design.

Particular acknowledgment is due to Associate Dean Elliott Axelband for writing Chapter 11 on the professionalization of systems architecting.

Acknowledgment is due from the authors to three members of the USC faculty, Dean Leonard Silverman, Associate Dean Richard Miller, now dean of engineering at the University of Iowa, and Professor Charles Weber of Electrical Engineering Systems, and to two members of the School of Engineering staff, Margery Berti and Mary Froehlig, who collectively created a Master of Science in Systems Architecture and Engineering out of an experimental course. They structured the program, helped recruit the students, relieved the authors of administrative concerns, and gave the program unquestioning dedication and loyalty. Most recently, Associate Dean Elliott Axelband has taken the baton as Eberhardt Rechtin's successor in teaching and directing the Masters program. Perhaps only other faculty members can appreciate how much such support means to a new field and its early authors.

To learn, teach — especially to industry-experienced, graduate-level students. Indeed, much has been learned by one of the authors of this book, Eberhardt Rechtin, in 6 years of actively teaching the concepts and processes of systems architecting to over 200 of them. At least a dozen of them deserve special recognition for their unique insights and penetrating commentary: Michael Asato, Kenneth Cureton, Susan Dawes, Norman P. Geis, Douglas R. King, Kathrin Kjos, Jonathan Losk, Ray Madachy, Archie W. Mills, Jerry Olivieri, Tom Pieronek, and Marilee Wheaton. The quick understanding and extension of the architecting process by all the students have been a joy to behold and are a privilege to acknowledge.

Following encouragement from Professor Weber in 1988 to take the first class offered in systems architecting as part of his Ph.D. in Electrical Engineering Systems, Mark Maier, its top student, went on to be an assistant professor at the University of Alabama in Huntsville and the coauthor of this book. He in turn thanks his colleagues at Hughes Aircraft who, years earlier, introduced him to system methods, especially Dr. Jock Rader, Dr. Kapriel Krikorian, and Robert Rosen.

The authors owe much to the work of Drs. Larry Druffel, David Garlan, and Mary Shaw of Carnegie Mellon University's Software Engineering Institute, Professor Barry Boehm of USC's Center for Software Engineering, and Dr. Robert Balzer of USC's Information Sciences Institute, software architects all, whose seminal ideas have migrated from their fields both to engineering systems in general and its architecting in particular.

Manuscripts may be written by authors, but publishing them is a profession and contribution unto itself, requiring judgment, encouragement, tact, and a considerable willingness to take risk. For all of these we thank Norm Stanton, senior editor at CRC Press, Inc., who has understood and supported the field, beginning with the publication of Frederick Brooks' classic architecting book, *The Mythical Man-Month*, two decades ago. Thanks are also due to Professor Stanley Weiss of MIT, Professor Terry Bahill of the University of Arizona, and to systems engineering consultant Ronald L. Wade for their thoughtful reviews of the manuscript, their corrections, suggestions, and, gratefully, their encouragement.

Eberhardt Rechtin
University of Southern California

Mark Maier
University of Alabama in Huntsville

The Authors

Dr. Rechtin recently retired from the University of Southern California (USC) as a professor with joint appointments in the departments of Electrical Engineering-Systems, Industrial & Systems Engineering, and Aerospace Engineering. His technical specialty is systems architecture in those fields. He had previously retired (1987) as President-emeritus of The Aerospace Corporation, an architect-engineering firm specializing in space systems for the United States Government.

Dr. Rechtin received his B.S. and Ph.D. degrees from CalTech in 1946 and 1950, respectively. He joined CalTech's Jet Propulsion Laboratory in 1948 as an engineer, leaving it in 1967 as an Assistant Director. At JPL he was the chief architect and director of the NASA/JPL Deep Space Network. In 1967 he joined the Office of the Secretary of Defense as the Director of the Advance Research Projects Agency and later as Assistant Secretary of Defense for Telecommunications. He left the Department of Defense in 1973 to become Chief Engineer of Hewlett-Packard. He was elected as the president and CEO of Aerospace in 1977, retiring in 1987, and joining USC to establish its graduate program in systems architecting.

Dr. Rechtin is a Member of the National Academy of Engineering and a Fellow of the Institute of Electrical and Electronic Engineers, the American Institute of Aeronautics and Astronautics, the American Association for Advancement of Science and is an Academician of the International Academy of Astronautics. He has been further honored by the IEEE with its Alexander Graham Bell Award, by the Department of Defense with its Distinguished Public Service Award, NASA with its Medal for Exceptional Scientific Achievement, the AIAA with its von Karman Lectureship, CalTech with its Distinguished Alumni Award, and by the (Japanese) NEC with its C&C Prize. Rechtin is also the author of *Systems Architecting, Creating and Building Complex Systems*, Prentice-Hall, 1991.

Dr. Maier graduated from CalTech with a bachelor's degree in electrical engineering in 1983 when he joined the Hughes Aircraft Company. At Hughes as a section head he was responsible for signal processing algorithm design and system engineering. While there he was awarded a Howard Hughes Doctoral Fellowship leading to a Ph.D. at USC. In 1992 he became an Assistant Professor of Electrical and Computer Engineering at the University of Alabama in Huntsville, where he is introducing architecting ideas in software engineering courses and doing research in digital stereo video compression and processing and signal processing for communications and radar. He is the author of several dozen journal and conference papers in these fields.

Dr. Maier began work in systems architecting at the very first 1989 seminar course on the subject and soon began research on its relationship with modeling methods and tools, with socio-technical systems, and with frameworks for organizing and evolving design heuristics. He is a consultant to firms on applying architecting methods to Intelligent Transport Systems and on the definition of enterprise product line.

Dedications

To Deedee, for whom life itself is an art.

Eberhardt Rechtin

To Leigh and Stuart, my anchors and inspiration.

Mark Maier

Contents

Chapter 2

Part two

Chapter 3

Part three

Chapter 7

Chapter 8

Chapter 9

Part four

Chapter 10

Chapter 11

Appendices

Part one

Introduction — the art of architecting

A brief review of classical architecting methods

Architecting: the art and science of designing and building systems.[1]

The four most important methodologies in the process of architecting are characterized as normative, rational, participative, and heuristic[2] (Table 1). As might be expected, like architecting itself, they contain both science and art. The science is largely contained in the first two, normative and rational, and the art in the last two, participative and heuristic.

Table 1   Four Architecting Methodologies

Normative (solution based)
Examples: building codes and communications standards
Rational (method based)
Examples: systems analysis and engineering
Participative (stakeholder based)
Examples: concurrent engineering and brainstorming
Heuristic (lessons learned)
Examples: **Simplify. Simplify. Simplify.** and **SCOPE!**

The normative technique is solution-based; it prescribes architecture as it "should be", that is, as given in handbooks, civil codes, and in pronouncements by acknowledged masters. Follow it and the result will be successful by definition.

Limitations of the normative method — such as responding to major changes in needs, preferences, or circumstances — led to the rational method and scientific and mathematical principles to be followed in arriving at a solution to a stated problem. It is method- or rule-based. Both the normative and rational methods are analytic, deductive, experiment-based, easily certified, well understood, and widely taught in academia and industry — truly a formidable set of positives.

However, although science-based methods are absolutely necessary parts of architecting, they are not the focus of this book. They are already well treated in a number of architectural and engineering texts. Equally necessary, and the focus of this part of the book, is the art, or practice, needed to complement the science for highly complex systems.

In contrast with science-based methodologies, the art or practice of architecting — like the practice of medicine, law, and business — is nonanalytic, inductive, difficult to certify, less understood, and, at least until recently, is seldom taught formally in either academia or industry. It is a process of insights, vision, intuitions, judgment calls, and even "taste".[3] It is key to creating truly new types of systems for new and often unprecedented applications. Here are some of the reasons.

For unprecedented systems, past data are of limited use. For others, analysis can be overwhelmed by too many unknowns, too many stakeholders, too many possibilities, and too little time for data gathering and analysis to be practical. To cap it off, many of the most important factors are not measurable. Perceptions of worth, safety, affordability, political acceptance, environmental impact, public health, and even national security provide no realistic basis for numerical analyses — even if they weren't highly variable and uncertain. Yet, if the system is to be successful, these perceptions must be accommodated from the first, top-level, conceptual model down through its derivatives.

The art of architecting, therefore, complements its science where science is weakest: in dealing with immeasurables, in reducing past experience and wisdom to practice, in conceptualization, in inspirationally putting disparate things together, in providing "sanity checks", and in warning of likely but unprovable trouble ahead. Terms like reasonable assumptions, guidelines, indicators, elegant design, and beautiful performance are not out of place in this art, nor are lemon, disaster, snafu, or loser — hardly quantifiable, but as real in impact as any science.

Participative methodology recognizes the complexities created by multiple stakeholders. Its objective is consensus. As a notable example, designers and manufacturers need to agree on a multiplicity of details if an end product is to be manufactured easily, quickly, and profitably. In simple but common cases, only the client, architect, and contractor have to be in agreement. But as systems become more complex, new and different participants have to agree as well.

Concurrent engineering, a recurrently popular acquisition method, was developed to help achieve consensus among many participants. Its greatest value, and its greatest contentions, are for systems in which widespread cooperation is essential for acceptance and success; for example, systems which directly impact on the survival of individuals or institutions. Its well-known weaknesses are undisciplined design by committee, diversionary brainstorming, the closed minds of "group think", and members without power to make decisions but with unbridled right to second guess. Arguably, the greatest mistake that can be made in concurrent engineering is to attempt to quantify it. It is not a science. It is a very human art.

The heuristics methodology is based on "common sense", that is, on what is sensible in a given context. Contextual sense comes from collective experience stated in as simple and concise a manner as possible. These statements are called heuristics, the subject of Chapter 2, and are of special importance to architecting because they provide guides through the rocks and shoals of intractable, "wicked" system problems. **Simplify!** is the first and probably most important of them. They exist in the hundreds if not thousands in architecting and engineering, yet they are some of the most practical and pragmatic tools in the architect's kit of tools.

Different methods for different phases of architecting

The nature of classical architecting changes as the project moves from phase to phase. In the earliest stages of a project it is a structuring of an unstructured mix of dreams, hopes, needs, and technical possibilities when what is most needed has been called an inspired synthesizing of feasible technologies. It is a time for the art of architecting. Later on, architecting becomes an integration of, and mediation among, competing subsystems and interests — a time for rational and normative methodology. And finally, there comes certification to all that the system is suitable for use, when it may take all the art and science to that point to declare the system as built is complete and ready for use.

Not surprisingly, architecting is often individualistic, and the end results reflect it. As Frederick P. Brooks put it in 1975[4] and Robert Spinrad stated in 1987,[5] the greatest architectures are the product of a single mind — or of a very small, carefully structured team — to which should be added in all fairness: a responsible and patient client, a dedicated builder, and talented designers and engineers.

Notes and References

1. *Webster's II New Riverside University Dictionary*, Riverside Publishing, Boston, 1984. As adapted for systems by substitution of "building systems" for "erecting buildings".
2. For a full discussion of these methods, see Lang, J., *Creating Architectural Theory, The Role of the Behavioral Sciences in Environmental Design*, Van Nostrand Reinhold, New York, 1987; and Rowe, P. G., *Design Thinking*, MIT Press, Cambridge, 1987. They are adapted for systems architecting in Rechtin, E., *Systems Architecting, Creating & Building Complex Systems*, Prentice-Hall, Englewood Cliffs, NJ, 1991, 14.

3. Spinrad, R. J., In a lecture to the University of Southern California, 1988.
4. Brooks, F. P., *The Mythical Man-Month, Essays on Software Engineering*, Addison-Wesley, Reading, MA, 1975 (and *Anniversary Edition*, 1995).
5. Spinrad, R. J., at a Systems Architecting lecture at the University of Southern California, fall 1987.

chapter one

Extending the architecting paradigm

Introduction: the classical architecting paradigm

The recorded history of classical architecting, the process of creating architectures, began in Egypt more than 4000 years ago with the pyramids, the complexity of which was overwhelming designers and builders alike. This complexity had at its roots the phenomenon that as systems became increasingly more ambitious, the number of inter-relationships among the elements increased far faster than the number of elements themselves. Pyramids were no longer simple burial sites; they had to be demonstrations of political and religious power, secure repositories of god-like rulers and their wealth, and impressive engineering accomplishments. Each demand, of itself, would require major resources. When taken together, they generated new levels of technical, financial, political, and social complications. Complex interrelationships among the combined elements were well beyond what the engineers' and builders' tools could handle.

And from that lack of tools for civil works came classical civil works architecture. Millennia later, technological advances in shipbuilding created the new and complementary fields of marine engineering and naval architecture. In this century, rapid advances in aerodynamics, chemistry, materials, electrical energy, communications, surveillance, information processing, and software have resulted in systems whose complexity is again overwhelming past methods and paradigms. One of those is the classical architecting paradigm.

Responding to complexity

> *Complex: composed of interconnected or interwoven parts.*[1]

> *System: a set of different elements so connected or related as to perform a unique function not performable by the elements alone.*[2]

It is generally agreed that increasing complexity* is at the heart of the most difficult problems facing today's systems of architecting and engineering. When architects and builders are asked to explain cost overruns and schedule delays, by far the most common, and quite valid, explanation is that the system is much more complex than originally thought. The greater the complexity, the greater the difficulty. It is important, therefore, to understand what is meant by system complexity if architectural progress is to be made in dealing with it.

The definitions of complexity and systems given at the start of this section are remarkably alike. Both speak to interrelationships (interconnections, interfaces, etc.) among parts or elements. As might be expected, the more elements and interconnections, the more complex the architecture and the more difficult the system-level problems.

Less apparent is that qualitatively different problem-solving techniques are required at high levels of complexity than at low ones. Purely analytical techniques, powerful for the lower levels, can be overwhelmed at the higher ones. At higher levels, architecting methods, experience-based heuristics, abstraction, and integrated modeling must be called into play.[3] The basic idea behind all of these techniques is to simplify problem solving by concentrating on its essentials. Consolidate and simplify the objectives. Stay within guidelines. Put to one side minor issues likely to be resolved by the resolution of major ones. Discard the nonessentials. Model (abstract) the system at as high a level as possible, then progressively reduce the level of abstraction. In short, **Simplify!**

It is important in reading about responses to complexity to understand that they apply throughout system development, not just to the conceptual phase. The concept that a complex system can be progressively partitioned into smaller and simpler units — and hence into simpler problems — omits an inherent characteristic of complexity, the interrelationships among the units. As a point of fact, poor aggregation and partitioning during development can *increase* complexity, a phenomenon all too apparent in the organization of work breakdown structures.

This primacy of complexity in system design helps explain why a single "optimum" seldom, if ever, exists for such systems. There are just too many variables. There are too many stakeholders and too many conflicting interests. No practical way may exist for obtaining information critical in making a "best" choice among quite different alternatives.

---

* A system need not be large or costly to be complex. The manufacture of a single mechanical part can require over 100 interrelated steps. A $10 microchip can contain thousands, even millions, of interconnected active elements.

The high rate of advances in the computer and information sciences

Unprecedented rates of advance in the computer and information sciences have further exacerbated an already complex picture. The advent of smart, software-intensive systems is producing a true paradigm shift in system design. Software, long treated as the glue that tied hardware elements together, is becoming the center of system design and operation. We see it in personal computers. The precipitous drop in computer hardware costs has generated a major design shift — from "keep the *computer* busy" to "keep the *user* busy." We see it in automobiles, where microchips increasingly determine the performance, quality, cost, and feel of cars and trucks.

We see the paradigm shift in the design of spacecraft and personal computers where complete character changes can be made in minutes. In effect, such software-intensive systems "change their minds" on demand. It is no longer a matter of debate whether machines have "intelligence"; the only real questions are of what kinds of intelligence and how best to use each one. And, because their software largely determines what and how the user perceives the system as a whole, its design will soon control and precede hardware design much as hardware design controls software today. This shift from "hardware first" to "software first" will force major changes on when and how system elements are designed — and who, with what expertise, will design the system as a whole. The impact on the value of systems to the user has been and will continue to be enormous.

Particularly susceptible to these changes are systems which depend upon electronics and information systems and which have not been characterized by the formal partnership with architecting that structural engineering has long enjoyed. This book is an effort to remedy that serious lack by extending the historical principles of classical architecting to modern information systems architecting.

The foundations of modern systems architecting

Although the day-to-day practice may differ significantly,[4] the foundations of modern systems architecting are much the same across many engineering disciplines. Generally speaking, they are a systems approach, a purpose orientation, a modeling methodology, ultraquality, certification, and insight.[5] Each will be described in turn.

A systems approach

A systems approach is one which focuses on the system as a whole, particularly when making value judgments (what is required) and design decisions (what is feasible). At the most fundamental level, *systems are collections of different things which together produce results unachievable by the elements alone.* For example, only when all elements are connected and working together do automobiles produce transportation, human organs produce life, and spacecrafts produce information. These system-produced results derive almost solely from the interrelationships among the elements, a fact that largely determines the technical role and principal responsibilities of the systems architect.

It is the responsibility of the architect to know and concentrate on the critical few details and interfaces that really matter and not to become overloaded with the rest. It is a responsibility that is important not only for the architect personally, but for effective relationships with the client and builder. To the extent that the architect is concerned with component design and construction, it is those details that specifically affect the system as a whole.

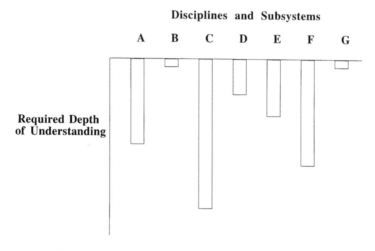

Figure 1.1 The architect's depth of understanding of subsystem and disciplinary details.

For example, a loaded question often posed by builders, project managers, and architecting students is, "How deeply should the architect delve into each discipline and each subsystem?" A graphic answer to that question is shown in Figure 1.1, exactly as sketched by Bob Spinrad in a 1987 lecture at the University of Southern California. The vertical axis is a relative measure of how deep into a discipline or subsystem an architect must delve to understand its consequences to the system as a whole. The horizontal axis lists the disciplines, such as electronics or stress mechanics, and/or the subsystems, such as computers or propulsion systems. Depending upon the system, a great deal or a very little depth of understanding may be necessary.

But that leads to another question, "How can the architect possibly know before there is a detailed system design, much less before system test, what details of what subsystem are critical?" A quick answer is — only through experience, through encouraging open dialog with subsystem specialists, and by being a quick, selective, tactful, and generally effective student of the system and its needs. Consequently and perhaps more than any other specialization, architecting is a continuing, day-to-day learning process. No two systems are exactly alike. Some are unprecedented, never built before.

> Exercise: Put yourself in the position of an architect asked to help a client build a system of a new type whose general nature you understand (a house, a spacecraft, a nuclear power plant, or a system in your own field) but which must considerably outperform an earlier version by a competitor. What do you expect to be the critical elements and details and in what disciplines or subsystems? What elements do you think you can safely leave to others? What do you need to learn the most about? Reminder: You will still be expected to be responsible for all aspects of the system design.

Critical details aside, the achitect's greatest concerns and leverage are still, and should be, with the systems' connections and interfaces: first, because they distinguish a system from its components; second, because their addition produces unique system-level functions, a primary interest of the systems architect, and third, because subsystem specialists are likely to concentrate most on the core and least on the periphery of their subsystems, viewing the latter as external constraints on their internal design. The specialists' concern for the system as a whole is understandably less than that of the systems architect; yet, if not managed well, the system functions can be in jeopardy.

A purpose orientation

Systems architecting is a process driven by a client's purpose or purposes. A President wants to meet an international challenge by safely sending astronauts to the moon and back. Military services need nearly undetectable strike aircraft. Cities call for pollutant-free transportation.

Clearly, if a system is to succeed, it must satisfy a useful purpose at an affordable cost for an acceptable period of time. Note the explicit value judgments in these criteria: a *useful* purpose, an *affordable* cost, and an *acceptable* period of time. Each and every one is the client's prerogative and responsibility, emphasizing the criticality of client participation in all phases of system acquisition. But of the three criteria, satisfying a useful purpose is predominant. Without it being satisfied, all others are irrelevant. Architecting therefore begins with, and is responsible for maintaining, the integrity of the system's purpose.

For example, the Apollo manned mission to the moon and back had a clear purpose, an agreed cost, and a no-later-than date. It delivered on all three. Those requirements, kept up front in every design decision, determined the mission profile of using an orbiter around the moon and not an earth-orbiting space station, and on developing electronics for a lunar orbit rendezvous instead of developing an outsized propulsion system for a direct approach to the lunar surface.

As another example, NASA headquarters, on request, gave the NASA/JPL Deep Space Network's huge ground antennas a clear set of priorities: first performance, then cost, then schedule even though the primary missions they supported were locked into the absolute timing of planetary arrivals. As a result, the first planetary communication systems were designed with an alternate mode of operation in case the antennas were not yet ready. As it turned out, and as a direct result of the NASA risk-taking decision, the antennas were carefully designed, not rushed, and not only satisfied all criteria for the first launch but for all launches for the next 30 years or so.

The Douglas Aircraft DC-3, though originally thought by the airline (now TWA) to require three engines, was rethought by the client and the designers in terms of its underlying purpose — to make a profit on providing affordable long distance air travel over the Rocky and Sierra Nevada mountains for paying commercial passengers. The result was the two-engine DC-3, the plane that introduced global air travel to the world.

In contrast, the promises of the Shuttle as an economically viable transporter to low earth orbit were far from met, although, in fact, the Shuttle did turn out to be a remarkably reliable and unique launch vehicle for selected missions. As of this date, the presumed purposes of the space station do not appear to be affordable; a change of purpose may be needed. The unacceptable cost/benefit ratios of the supersonic transport, the space-based ballistic missile defense system, and the superconducting supercollider terminated all these projects before their completion.

Curiously, the end use of a system is not always what was originally proposed as its purpose. The F-16 fighter aircraft was designed for visual air-to-air combat but most used for ground support. The ARPANET-INTERNET communication network originated as a government-furnished computer-to-computer linkage in support of university research; it is now most used, and paid for, by individuals for e-mail. Both are judged as successful. Why? Because, as circumstances changed, providers and users redefined the meaning of useful, affordable, and acceptable. A useful heuristic comes to mind: **Design the structure with "good bones".** It comes from the architecting of buildings, bridges, and ships where it refers to structures that are resilient to a wide range of stresses and changes in purpose. It could just as well come from physiology and the remarkably adaptable spinal column and appendages of all verte-brates — fishes, amphibians, reptiles, birds, and mammals.

> Exercise: Identify a system whose purpose is clear and unmistak-able. Identify, contact, and, if possible, visit its architect. Compare notes and document what you learned.

Technology-driven systems, in notable contrast to purpose-driven sys-tems, tell a still different story. They are the subject of Chapter 3, Builder-Architected Systems.

A modeling methodology

Modeling is the creation of abstractions or representations of the system to predict and analyze performance, costs, schedules, and risks and to provide guidelines for systems research, development, design, manu-facture, and management. Modeling is the centerpiece of systems archi-tecting — a mechanism of communication to clients and builders, of design management with engineers and designers, of maintaining sys-tem integrity with project management, and of learning for the archi-tect, personally.

> Examples: The balsa wood and paper scale models of a residence, the full-scale mockup of a lunar lander, the rapid prototype of a software application, the computer model of a communication net-work, or the mental model of a user.

Modeling is of such importance to architecting that it is the sole subject of Part Three. At this point it is important to clear up a misconception. There has been a common perception that modeling is solely a *concep-tual* process; that is, its purpose is limited to representing or simulating only the general features of an eventual system. If so, a logical question can be raised: when to stop modeling and start building. (It could almost have been asked as: when to stop architecting and begin engi-neering!)

Indeed, just that question was posed by Mark Maier in the first USC graduate class. Thinking had just begun on the so-called "stop question" when it was realized that modeling did *not* stop at the end of conceptualization in any case. Rather it progressed, evolved, and solved problems from the beginning of a system's acquisition to its final retirement. There are, of course, conceptual models, but there are also engineering models and subsystem models; models for simulation, prototypes, and system test; demonstration models, operational mod-els, and mental models by the user of how the system behaves.

Models are, in fact, created by many participants, not just by archi-tects. These models must somehow be made consistent with overall system imperatives. It is particularly important that they be consistent with the architect's system model, a model that evolves, becoming more and more concrete and specific as the system is built. It provides a standard against which consistency can be maintained and is a powerful tool in maintaining the larger objective of system integrity. And finally, when the system is operational and a deficiency or failure appears, a model — or full-scale simulator if one exists — is brought into play to help determine the causes and cures of the problem. The more complete the model, the more accurately possible failure mechanisms can be duplicated until the one and only cause is identified.

In brief, modeling is a multipurpose, progressive activity, evolving and becoming less abstract and more concrete as the system is built and used.

Ultraquality implementation

Ultraquality is defined as a level of quality so demanding that it is impractical to measure defects, much less certify the system prior to use.[6] It is a limiting case of quality driven to an extreme, a state beyond acceptable quality limits (AQL) and statistical quality control. It requires a zero defect approach not only to manufacturing, but to design, engineering, assembly, test, operation, maintenance, adaptation, and retirement — in effect, the complete life cycle.

Some examples are a new-technology spacecraft with a design lifetime of at least 10 years, a nuclear power plant that will not fail catastrophically within the foreseeable future, and a communication network of millions of nodes, each requiring almost 100% availability. Ultraquality is a recognition that the more components there are in a system, the more reliable each component must be to a point where, at the element level, defects become impractical to measure within the time and resources available. Yet, the reliability goal of the system as a whole must still be met. In effect, it reflects the not unreasonable demand that a system — regardless of size or complexity — should not fail to perform more than about 1% — or less — of the time. An ICBM shouldn't. A Shuttle, at least 100 times more complex, shouldn't. An automobile shouldn't. A passenger airliner, at least 100 times more complex, shouldn't; as a matter of fact, we expect the airliner to fail far, far less than the family car.

> Exercise: Trace the histories of commercial aircraft and passenger buses over the last 50 years in terms of the number of trips that a passenger would expect to make without an accident. What does that mean to vehicle reliability as trips lengthen and become more frequent, as vehicles get larger and more complex? How were today's results achieved? What trends do you expect in the future?

The subject would be moot if it were not for the implications of this "limit state" of zero defects to design. Zero defects, in fact, originated as long ago as World War II, largely driven by patriotism. As a motivator, the zero defects principle was a prime reason for the success of the Apollo mission to the moon.

To show the implications of ultraquality processes, if a manufacturing line operated with zero defects there would be no need, indeed it would be worthless, to build elaborate instrumentation and information processing support systems. This would reduce costs and time instead by 30%. If an automobile had virtually no design or production defects, then sales outlets would have much less need for large service shops with their high capital and labor costs. Indeed, the service departments of the finest automobile manufacturers are seldom fully booked, resembling something like the famous Maytag commercial. Very little repair or service, except for routine maintenance, is required for 50 to 100,000 miles. Not coincidentally, these shops invariably are spotlessly clean, evidence of both the professional pride and discipline required for sustaining an ultraquality operation. Conversely, a dirty shop floor is one of the first and most telling indicators to a visitor or inspector of low productivity, careless workmanship, reduced plant yield, and poor product performance. The rocket, ammunition, solid-state component, and automotive domains all bear witness to that fact.

As another example, microprocessor design and development has maintained the same per-chip defect rate even as the number and complexity of operations increased by factors of thousands. The corresponding failure rate per individual operation is now so low as to be almost unmeasurable. Indeed, for personal computer applications, a microprocessor hardware failure once a year borders on the unacceptable.

Demonstrating this limit state in high quality is not a simple extension of existing quality measures, though the latter may be necessary in order to get within range of it. In the latter there is an heuristic: (measureable) **acceptance tests must be both complete and passable**. How then can inherently unmeasurable ultraquality be demanded or certified? Most of the answer seems to be in surrogate procedures, such as zero defects program, and less in measurements, not because finer measurements would not be useful, but because system complexity has outpaced instrument accuracy.

In that respect, a powerful addition to pre-1990 ultraquality techniques was the concept, introduced in the last few years, that each participant in a system acquisition sequence is both a buyer and a supplier. The original application, apparently a Japanese idea, was that each worker on a production line was a buyer from the preceding worker in the production line as well as a supplier to the next. Each role required a demand for high quality, that is, a refusal to buy a defective item and a concern not to deliver a defective one likely to be

refused.[7] In effect, the supplier–buyer concept generates a self-enforc-
ing quality program with built-in inspection. There would seem to be
no reason why the same concept should not apply throughout system
acquisition — from architect to engineer to designer to producer to
seller to end user. As with all obvious ideas, the wonder is why it wasn't
self-evident earlier.

Certification

Certification is a formal statement by the architect to the client or user
that the system, as built, meets the criteria both for client acceptance
and for builder receipt of payment. Certification is the grade on the
"final exams" of system test and evaluation. To be accepted it must be
well supported, objective, and fair to client and builder alike.

> Exercise: Pick a system for which the purposes are reasonably
> clear. What tests would you, as a client, demand be passed for you
> to accept and pay for the system? What tests would you, as a
> builder, contract to pass in order to be paid? Whose word would
> each of you accept that the tests had or had not been passed?

Clearly, if certification is to be unchallenged, then there must be no
perception of conflict of interest of the architect. This imperative has
led to three widely accepted, professionally understood constraints[8] on
the role of the architect:

1. *A disciplined avoidance of value judgments,* that is, of intruding in
   questions of worth to the client; questions of what is satisfactory,
   what is acceptable, affordable, maintainable, reliable, and so on.
   Those judgments are the imperatives, rights, and responsibilities
   of the client. As a matter of principle, the client should judge on
   desirability and the architect should decide (only) on feasibility.
   To a client's question of "What would you do in my position?"
   the experienced architect responds only with further questions
   until the client can answer the original one. To do otherwise
   makes the architect an advocate and, in some sense, the "owner"
   of the end system, preempting the rights and responsibilities of
   the client. It may make the architect famous, but the client will
   feel used. Residences, satellites, and personal computers have
   all suffered from such preemption.
2. *A clear avoidance of perceived conflict of interest* through participation in research and development, including ownership or
   participation in organizations that can be, or are, building the
   system. The most evident conflict here is the architect recom-
   mending a system element that the architect will supply and

profit from. This constraint is particularly important in builder-architected systems (Chapter 3).*

3. *An arms-length relationship with project management,* that is, with the management of human and financial resources other than of the architect's own staff. The primary reason for this arrangement is the overload and distraction of the architect created by the time-consuming responsibilities of project management. A second conflict, similar to that of participating in research and development, is created whenever architects give project work to themselves. If clients, for reasons of their own, nonetheless ask the architect to provide project management, it should be considered as a separate contract for a different task requiring different resources.

Insight and heuristics

A picture is worth a thousand words.

Chinese Proverb, 1000 B.C.

One insight is worth a thousand analyses.

Charles Sooter, April 1993

Insight, or the ability to structure a complex situation in a way that greatly increases understanding of it, is strongly guided by lessons learned from one's own or others' experiences and observations. Given enough lessons, their meaning can be codified into succinct expressions called "heuristics", a Greek term for guide. Heuristics are an essential complement to analytics, particularly in situations where analysis alone cannot provide either insights or guidelines.[9] In many ways they resemble what are called principles in other arts, for example, the importance of balance and proportion in a painting, a musical composition, or the ensemble of a string quartet. Whether as heuristics or principles, they encapsulate the insights that have to be achieved and practiced before a masterwork can be achieved.

Both architecting and the fine arts clearly require insight and inspiration as well as extraordinary skill to reach the highest levels of achievement. Seen from this perspective, the best systems architects are indeed artists in what they do. Some are even artists in their own right. Renaissance architects like Michaelangelo and Leonardo da Vinci were also consumate artists. They not only designed cathedrals, they

---

* Precisely this constraint led the Congress to mandate the formation in 1960 of a nonprofit engineering company, The Aerospace Corporation, out of the for-profit TRW Corporation, a builder in the aerospace business.

executed the magnificent paintings in them. The finest engineers and architects, past and present, are often musicians; Simon Ramo and Ivan Getting, famous in the missile and space field and, respectively, a violinist and pianist, are modern-day examples.

The wisdom that distinguishes the great architect from the rest is that insight and inspiration, which, combined with well-chosen methods and guidelines and fortunate circumstances, create masterworks. Unfortunately, wisdom does not come easily. As one conundrum puts it:

> **Success comes from wisdom.**
> **Wisdom comes from experience.**
> **Experience comes from mistakes.**

Therefore, because success comes only after many mistakes, something few clients would willingly support, it presumably is either unlikely or must follow a series of disasters.

This reasoning might well apply to an individual. But applied to the profession as a whole, it clearly does not. The required mistakes — and experience and wisdom gained from them — can be those of one's predecessors, not necessarily one's own.

And from that understanding comes the role of education. It is the place of education to research, document, organize, codify, and teach those lessons so that mistakes need not be repeated as a prerequisite for future successes. Chapter 2, heuristics as tools, is a start in that direction for the art of systems architecting.

However, before exploring the creation and use of heuristics it is necessary to place systems architecting in context.

The waterfall model of systems acquisition

As with systems, no process exists by itself. All processes are part of still larger ones; and all processes have subprocesses. The process of architecting is no exception.

Architecting exists within a still larger process that includes design, engineering, manufacturing, testing, and operation, that is, the process of systems acquistion. This larger process is classically depicted as an expanded waterfall (Figure 1.2).[10] The architect's functional relationship with this larger process is sketched in Figure 1.3. Managerially, the architect could be a member of the client's or the builder's organization, or of an independent architecting partnership, as long as perceptions of conflict of interest are avoided. In any case and wherever the architect is physically or managerially located, the relationships to the client and the acquisition process are essentially as shown. The strongest (thickest line) decision ties are with client need and resources, conception and

model building, and with testing, certification, and acceptance. Less
prominent are the monitoring ties with engineering and manufacturing. There are also important, if indirect, ties with social and political
factors, the "illities", and the "real world".

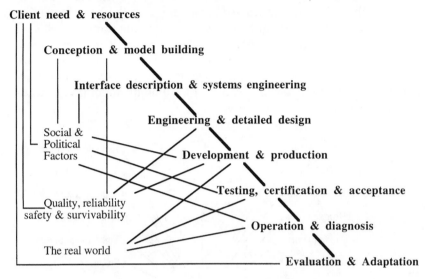

Figure 1.2   The expanded waterfall.

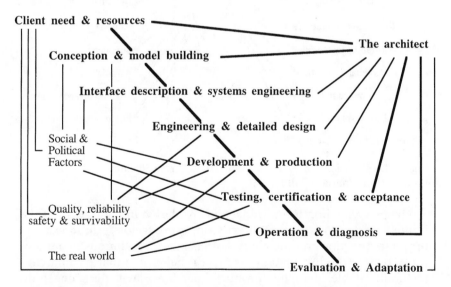

Figure 1.3   The architect and the expanded waterfall. (Adapted from Rechtin
1991. With permission from Prentice-Hall.)

This waterfall model of systems acquisition has served systems acquistion well for centuries, evolving as new technologies create new, larger-scale, more complex systems of all types. The most recent qualitative changes are due to the needs of software-intensive systems, as will be seen in Chapters 4 and 6 and in Part 3. While these changes increase the role and importance of architecting, the basic functional relationships shown in Figure 4.3 remain much the same.

These relationships are more complex than simple lines might suggest. As well as indicating channels for two-way communication and upward reporting, they infer the tensions to be expected between the connected elements, tensions caused by different imperatives, needs, and perceptions.

Figure 1.4   Tensions in systems architecting. (From Rechtin 1991. With permission from Prentice-Hall.)

Some competing technical factors are shown in Figure 1.4.[11] This figure was drawn such that directly opposing factors pull in exactly opposite directions on the chart. For example, continuous evolution pulls against product stability; a typical balance is that of an architecturally stable, evolving product line. Low level decisions pull against strict process control, which can often be relieved by systems architectural partitioning, aggregation, and monitoring. Most of these trade-offs can be expressed in analytic terms, which certainly helps, but some cannot, as will become apparent in the social systems world of Chapter 5.

> Exercise: Give examples from a specific system of what information, decisions, recommendations, tasks, and tensions might be expected across the lines of Figure 4.3.

Summary and conclusions

A system is a collection of different things which together produce results unachievable by themselves alone. The value added by systems is in the interrelationships of their elements.

Architecting is creating and building structures, i.e., "structuring". *Systems* architecting is creating and building *systems*. It strives for fit, balance, and compromise among the tensions of client needs and resources, technology, and multiple stakeholder interests.

Architecting is both an art and a science — both synthesis and analysis, induction and deduction, and conceptualization and certification — using guidelines from its art and methods from its science. As a process, it is distinguished from systems engineering in its greater use of heuristic reasoning, lesser use of analytics, closer ties to the client, and particular concern with certification of readiness for use.

The foundations of systems architecting are a systems approach, a purpose orientation, a modeling methodology, ultraquality, certification, and insight. To avoid perceptions of conflict of interest, architects must avoid value judgments, avoid perceived conflicts of interest, and keep an arms-length relationship with project management.

A great architect must be as skilled as an engineer and as creative as an artist or the work will be incomplete. Gaining the necessary skills and insights depends heavily on lessons learned by others, a task of education to research and teach.

The role of systems architecting in the systems acquisition process depends upon the phase of that process. It is strongest during conceptualization and certification, but never absent. Omitting it at any point, as with any part of the acquisition process, leads to predictable errors of omission at that point to those connected with it.

Notes and References

1. *Webster II, New Riverside University Dictionary,* Riverside Publishing, Boston, 1984, 291.
2. Rechtin 1991, p. 7.

3. Another factor in overruns and delays is uncertainty, the so-called un-knowns and unknown unknowns. Uncertainty is highest during con-ceptualization, less in design, still less in redesign, and least in an upgrade. As with complexity, the higher the level, the more important become experience-based architecting methods. See Carpenter, R. G., *System Architects' Job Characteristics and Approach to the Conceptualization of Complex Systems.* Doctoral dissertation in Industrial and Systems En-gineering, University of Southern California, Los Angeles, 90089-1450, 1995.
4. Cuff, D., *Architecture: The Story of Practice*, MIT Press, Cambridge, 1991.
5. Rechtin, E., Foundations of systems architecting, *Syst. Eng. J. Nat. Counc. Syst. Eng.*, 1(1), 35, July/September 1994.
6. Discussed extensively in Juran, J. M., *Juran on Planning for Quality*, Free Press, New York, 1988; Phadke, M. S., *Quality Engineering Using Robust Design*, Prentice-Hall, Englewood Cliffs, NJ, 1989; and Rechtin, E., *Sys-tems Architecting, Creating & Building Complex Systems*, Prentice-Hall, Englewood Cliffs, NJ, 1991, 160.
7. See also Chapter 4, Manufacturing Systems.
8. From Cantry, D., What architects do and how to pay them, *Arch. Forum*, 119, 92, September 1963; and from discussions with Roland Russell, AIA.
9. See King, D. R., The Role of Informality in System Development or, A Critique of Pure Formality, University of Southern California, 1992 (un-published but available through USC).
10. Rechtin, E., *Systems Architecting, Creating & Building Complex Systems*, Prentice-Hall, Englewood Cliffs, NJ, 1991, 4.
11. Rechtin, E., *Systems Architecting, Creating & Building Complex Systems*, Prentice-Hall, Englewood Cliffs, NJ, 1991, 156.

chapter two

Heuristics as tools

Introduction: a metaphor

Mathematicians are still smiling over a gentle self-introduction by one of their famed members. "There are *three* kinds of mathematicians," he said, "those that know how to count and those that don't." The audience waited in vain for the third kind until, with laughter and appreciation, they caught on. Either the member couldn't count to three — ridiculous — or he was someone who believed that there was more to mathematics than numbers, important as they were. The number theorists appreciated his acknowledgment of them. The "those that don'ts" quickly recognized him as one of their own, the likes of a Gödel who, using thought processes alone, showed that no set of theorems can ever be complete.

Modifying the self-introduction only slightly to the context of this chapter, there are three kinds of people in our business, those that know how to count and those that don't — including the authors.

Those that know how to count (most engineers) approach their design problems using analysis and optimization, powerful and precise tools derived from the scientific method and calculus. Those that don't (most architects) approach their qualitative problems using guidelines, abstractions, and pragmatics generated by lessons learned from experience, that is, heuristics. As might be expected, the tools each use are different because the kinds of problems they solve are different. We routinely and accurately describe an individual as "thinking like an engineer" — or architect, or scientist, or artist. Indeed, by their tools and works ye shall know them.

This chapter, metaphorically, is about architects' heuristic tools. As with the tools of carpenters, painters, and sculptors, there are literally hundreds of them — but only a few are needed at any one time and for a specific job at hand. To continue the metaphor, although a few tool users make their own, the best source is usually a tool supply store — whether it be for hardware, artists' supplies, software — or heuristics. Appendix A, Heuristics for Systems-Level Architecting, is an heuristics store, organized by task, just like any good hardware store.

Customers first browse, then select a kit of tools based on the job, personal skill, and a knowledge of the origin and intended use of each tool.

Heuristic has a Greek origin, *heuriskein,* a word meaning "to find a way" or "to guide" in the sense of piloting a boat through treacherous shoals. Architecting is a form of piloting. Its rocks and shoals are the risks and changes of technology, construction, and operational environment that characterize complex systems. Its safe harbors are client acceptance and safe, dependable, long life. Heuristics are guides along the way — channel markings, direction signs, alerts, warnings, and anchorages — tools in the larger sense. But they must be used with judgment. No two harbors are alike. The guides may not guarantee safe passage, but to ignore them may be fatal. The stakes in architecting are just as high — reputations, resources, vital services, and, yes, lives. Consonant with their origin, the heuristics in this book are intended to be trusted, time-tested guidelines for serious problem solving.

Heuristics as so defined are narrower in scope, subject to more critical test and selection, and intended for more serious use than other guidelines; for example conventional wisdom, aphorisms, maxims, rules of thumb, and the like. To illustrate: a pair of mutually contradictory statements like (1) *look before you leap* and (2) *he who hesitates is lost* are hardly useful guides when encountering a cliff while running for your life. In this book, neither of this pair would be a valid heuristic because they offer contradictory advice for the same problem.

The purpose of this chapter is therefore to help the reader — whether architect, engineer, or manager — find or develop heuristics that can be trusted, organize them according to need, and use them in practice. The first step is to understand that heuristics are abstractions of experience.

Heuristics as abstractions of experience

One of the most remarkable characteristics of the human race is its ability not only to learn, but to pass on to future generations sophisticated abstractions of lessons learned from experience. Each generation knows more, learns more, plans more, tries more, and succeeds more than the previous one because it needn't repeat the time-consuming process of reliving the prior experiences. Think of how extraordinarily efficient are such quantifiable abstractions as $F = ma$, $E = mc^2$, and $x = F(y,z,t)$; of algorithms, charts, and graphs; and of the basic principles of economics. This kind of efficiency is essential if large, lengthy, complex systems and long-lived product lines are to succeed. Few architects ever work on more than two or three of them in a lifetime. They have neither the time nor opportunity to gain the experience needed to create first-rate architectures from scratch. By much the same process qualitative heuristics, condensed and codifed practical experience, came into being to complement the equations and algorithms of science and engineering in the solving of complex problems. Passed from architect to architect, from system to system, they worked. They helped satisfy a real need.

In contrast to the symbols of physics and mathematics, the format of heuristics is words expressed in the natural languages. Unavoidably they reflect the cultures of engineering, business, exploration, and human relations in which they arose. The birth of an heuristic begins with anecdotes and stories, hundreds of them, in many fields; which become parables, fables, and myths,[1] easily remembered for the lessons they teach. Their impact, even at this early stage, can be remarkable not only on politics, religion, and business, but on the design of technical systems and services. The lessons that have endured are those that have been found to apply beyond the original context, extended there by analogy, comparison, conjecture, and testing.* At their strongest they are seen as self-evident truths requiring no proof.

There is an interesting human test for a good heuristic. An experienced listener, on first hearing one, will know within seconds that it fits that individual's model of the world. Without having said a word to the speaker, the listener almost invariably affirms its validity by an unconscious nod of the head, and then proceeds to recount a personal experience that strengthens it. Such is the power of the human mind.

Selecting a personal kit of heuristic tools

The art in architecting lies not in the wisdom of the heuristics, but in the wisdom of knowing which heuristics apply, a priori, to the current project.[2]

All professions and their practitioners have their own kits of tools, physical and heuristic, selected from their own and others' experiences to match their needs and talents. But, in the case of architecting and engineering prior to the late 1980s, selections were limited and, at best, difficult to acquire. An effort was therefore made in the USC graduate course in Systems Architecting to create a much wider selection by gathering together lessons learned throughout the West Coast aerospace, electronics, and software industries and expressing them in heuristic form for use by architects, educators, researchers, and students.

---

* This process is one of inductive reasoning, "a process of truth estimation in the face of incomplete knowledge which blends information known from experience with plausible conjecture." Klir, G. J., *Architecture of Systems Problem Solving*, Plenum Press, New York, 1985, 275. More simply, it is an extension or generalization from specific examples. It contrasts with deductive reasoning which derives solutions for specific cases from general principles.

An initial collection[3] of about 100 heuristics was soon surpassed by contributions from over 200 students, reaching nearly 1000 heuristics within 6 years.[4] Many, of course, were variations on single, central ideas — just as there are many variations of hammers, saws, and screwdrivers — repeated time and time again in different contexts. The four most widely applicable of these heuristics were, in decreasing order of popularity:

- **Don't assume that the original statement of the problem is necessarily the best, or even the right one.**

Example: The original statement of the problem for the F-16 fighter aircraft asked for a high-supersonic capability, difficult and expensive to produce. Discussions with the architect, Harry Hillaker, brought out that the reason for this statement was to provide a quick exit from combat, something far better provided by a high thrust-to-weight, low supersonic design. In short, the original high-speed statement was replaced by a high acceleration one, with the added advantage of exceptional maneuverability.

- **In partitioning, choose the elements so that they are as independent as possible; that is, elements with low external complexity and high internal complexity.**

Example: One of the difficult problems in the design of microchips is the efficient use of their surface area. Much of that area is consumed by connections between components, that is, by communications rather than by processing. Professor Carver Mead of CalTech has now demonstrated that a design based on minimum communications between process-intensive nodes results in much more efficient use of space, with the interesting further result that the chip "looks elegant" — a sure sign of a fine architecture and another confirmation of the heuristic, **The eye is a fine architect. Believe it.**

- **Simplify. Simplify. Simplify.**

Example: One of the best techniques for increasing reliability while decreasing cost and time is to reduce the piece part count of a device. Automotive engineers, particularly recently, have produced remarkable results by substituting single castings for multiple assemblies and by reducing the number of fasteners and their associated assembly difficulties by better placement. A comparable result in computer software is the use of reduced instruction set computers (RISC) which increases computing speed for much the same hardware.

- **Build in and maintain options as long as possible in the design and implementation of complex systems. You will need them.**

Example: In the aircraft business they are called "scars". In the software business they are called "hooks". Both are planned breaks or entry points into a system which can extend the functions the system can provide. For aircraft, they are used for lengthening the fuselage to carry more passengers or freight. For software, they are used for inserting further routines.

Though these four heuristics do not make for a complete tool kit, they do provide good examples for building one. All are aimed at reducing complexity, a prime objective of systems architecting. All have been trusted in one form or another in more than one domain. All have stood the test of time for decades if not centuries.

The first step in creating a larger kit of heuristics is to determine the criteria for selection. The following were established to eliminate unsubstantiated assertions, personal opinions, corporate dogma, anecdotal speculation, mutually contradictory statements, and the like. As it turned out, they also helped generalize domain-specific heuristics into more broadly applicable statements. The strongest heuristics passed all the screens easily. The criteria were

- The heuristic must make sense in its original domain or context. To be accepted, a strong correlation, if not a direct cause and effect, must be apparent between the heuristic and the successes or failures of specific systems, products, or processes. Academically speaking, both the rationale for the heuristic and the report that provided it were subject to peer and expert review. As might be expected, a valid heuristic seldom came from a poor report.
- The general sense, if not the specific words, of the heuristic should apply beyond the original context. That is, the heuristic should be useful in solving or explaining more than the original problem from which it arose. An example is the foregoing **don't assume** heuristic. Another is **Before the flight it's opinion; after the flight it's obvious**. In the latter, the word "flight" can be sensibly replaced by test, experiment, fight, election, proof, or trial. In any case, the heuristic should not be wrong or contradictory in other domains where it could lead to serious misunderstanding and error.
- The heuristic should be easily rationalized in a few minutes or on less than a page. As one of the heuristics itself states, "**If you can't explain it in five minutes, either you don't understand it or it doesn't work** (Darcy McGinn, 1992, from David Jones). With that in mind, the more obvious the heuristic is on its face,

and the fewer the limitations on its use, the better. Example: **A model is not reality.**

- The opposite statement of the heuristic should be foolish, clearly not "comon sense". For example, the opposite of **Murphy's Law** —**"If it can fail, it will"**— would be "If it can fail, it won't", which is patent nonsense.
- The heuristic's lesson, though not necessarily its most recent formulation, should have stood the test of time and earned a broad consensus. Originally this criterion was that the heuristic *itself* had stood the test of time, a criterion that would have rejected recently formulated heuristics based on retrospective understanding of older or lengthy projects.

Example: **The beginning is the most important part of the work** (Plato, 4th century B.C.), reformulated more recently as **All the serious mistakes are made in the first day** (Robert Spinrad, 1988).

It is probably true that heuristics can be even more useful if they can be used in a set, like wrenches and screwdrivers, hammers and anvils, or files and vises. The taxonomy grouping to follow achieves that possibility in part.

It is also probably true that a proposed action or decision is stronger if it is consistent with several heuristics rather than only one. A set of heuristics applicable to acceptance procedures substantiates that proposition.

And it would certainly seem desirable that an heuristic, taken in a sufficiently restricted context, could be specialized into a design rule, a quantifiable, rational evaluation, or a decision algorithm. If so, heuristics of this type would be useful bridges between architecting, engineering, and design.

Heuristics on heuristics

A phenomenon observed as heuristics was discovered by the USC graduate students, which is that the discoverers themselves began thinking heuristically. They found themselves creating heuristics directly from observation and discussion, then trying them out on professional architects and engineers, some of whose experiences had suggested them. (Most interviewees were surprised and pleased at the results.) The resultant provisional heuristics were then submitted for academic review as parts of class assignments.

Kenneth L. Cureton, carrying the process one step further, generated a set of heuristics on how to generate and apply heuristics,[5] from which the following were chosen.

Generating useful heuristics

- Humor (and careful choice of words) in a heuristic provide an emotional bite that enhances the mnemonic effect. (Karklins)
- Use words that transmit the "thrill of insight" into the mind of the beholder.
- For maximum effect, try embedding both descriptive and pre-scriptive messages in a heuristic.
- Many heuristics can be applied to heuristics themselves (e.g., **Simplify! Scope!**).
- Don't make a heuristic so elegant that it only has meaning to its creator, and thus loses general usefulness.
- Rather than adding a conditional statement to a heuristic, con-sider creating a separate but associated heuristic that focuses on the insight of dealing with that conditional situation.

Applying heuristics

- If it works, then it's useful.
- Knowing when and how to use a heuristic is as important as knowing the what and why.
- Heuristics work best when applied early to reduce the solution space.
- Strive for balance — too much of a good thing or complete elimination of a bad thing may make things worse, not better!
- Practice — practice — practice!
- Heuristics aren't reality, either!

A taxonomy of heuristics

The second step after finding or creating individual heuristics is to organize them for easy access so that the appropriate ones are at hand for the immediate task. The collection mentioned earlier in this chapter was accordingly refined and organized by architecting task.* In some ways, the resultant list — presented in Appendix A — was self-orga-nizing. Heuristics tended to cluster around what became recognized as basic architecting tasks. For example, although certifying is shown last and is one of the last formal phases in a waterfall, it actually occurs at many milestones as "sanity checks" are made along the way and

---

* The original 100 of Rechtin 1991 were organzed by the phases of a waterfall. The list in Appendix A of this book recognizes that many heuristics apply to several phases, that the spiral model of system development would in any case call for a different categori-zation, and that many of the tasks described here occur over and over again during systems development.

subsystems are assembled. The tasks, elaborated in Chapter 8, p. 145 to 152, are

- Scoping and planning
- Modeling
- Prioritizing
- Aggregating
- Partitioning
- Integrating
- Certifying
- Assessing
- Evolving and rearchitecting

The list is further refined by distinguishing between two forms of heuristic. One form is descriptive; that is, it describes a situation but does not indicate directly what to do about it. Another is prescriptive; that is, it prescribes what might be done about the situation. An effort has been made in the appendix to group prescriptions under appropriate descriptions with some, but not complete, success. Even so, there are more than enough generally applicable heuristics for the reader to get started.

And then there are sets of heuristics that are domain specific to aircraft, spacecraft, software, manufacturing, social systems, and so on. Some of these can be deduced or specialized from more general ones given here. Or, they can be induced or generalized from multiple examples in specialized subdomains. Still more fields are explored in Part 3, adding further heuristics to the general list.

The readers are encouraged to discover still more, general and specialized, in much the same way the more general ones here were — by spotting them in technical journals, books,[6] project reports, management treatises, and conversations.

The Appendix A taxonomy is not the only possible one any more than all tool stores are organized in the same way. In Appendix A one heuristic follows another, one-dimensionally, as in any list. But some are connected to others in different categories, or could just as easily be placed there. Some are "close" to others and some are further away. Dr. Ray Madachy, then a graduate student, using a form of object-oriented programming, converted the list into a two-dimensional, interconnected "map" in which the main nodes were architecting themes: conception and design; the systems approach; quality and safety; integration, test, and certification; and disciplines.[7] To these were linked each of the 100 heuristics in the first systems architecting text,[8] which in turn were linked to each other. The ratio of heuristic-to-heuristic links to total links was about 0.2; that is, about 20% of the heuristics overlapped into other nodes.

The Madachy taxonomy, however, shared a limitation common to all object-oriented methods — the lack of upward scalability into hundreds of objects — and consequently was not used for Appendix A. Nonetheless, it could be useful for organizing a modest-sized personal tool kit or for treating problems already posed in object-oriented form, for example, computer-aided design of spacecraft.[9]

Summary

Heuristics, as abstractions of experience, are trusted, nonanalytic guidelines for treating inherently unbounded, ill-structured problems. They are used as aids to decision making, value judgments, and assessments. They are found throughout systems architecting, from earliest conceptualization through diagnosis and operation. They provide bridges between client and builder, concept and implementation, synthesis and analysis, and system and subsystem. They provide the successive transitions from qualitative, provisional needs to descriptive and prescriptive guidelines, and thence to rational approaches and methods.

This chapter introduces the concept of heuristics as tools — how to find, create, organize, and use them for treating the qualitative problems of systems architecting. Appendix A provides a ready source of them organized by architecting task; in effect, a tool store of systems architecting heuristic tools.

Notes and References

1. For more of the philosophical basis of heuristics, see Asato, M., The Power of the Heuristic, USC, 1988; and Rowe, A. J. , The meta logic of cognitively based heuristics, in *Expert Systems in Business and Finance: Issues and Applications*, Watkins, P. R. and Eliot, L. B., Eds., John Wiley & Sons, New York, 1988.
2. Williams, P. L., 1992 Systems Architecting Report, University of Southern California, unpublished.
3. Rechtin, E., *Systems Architecting, Creating & Building Complex Systems*, Prentice-Hall, Englewood Cliffs, NJ, 1991, 312.
4. Rechtin, E., Ed., Collection of Student Heuristics in Systems Architecting, 1988–93, University of Southern California, unpublished, 1994.
5. Cureton, K. L., Metaheuristics, USC Graduate Report, December 9, 1991.
6. Many of the ones in Appendix A come from a similar appendix in Rechtin, E., *Systems Architecting, Creating & Building Complex Systems*, Prentice-Hall, Englewood Cliffs, NJ, 1991, Appendix.
7. Madachy, R., Thread Map of Architecting Heuristics, USC 4/21/91; and Formulating Systems Architecting Heuristics for Hypertext, USC, April 29, 1991.

8. Rechtin, E., *Systems Architecting, Creating & Building Complex Systems,* Prentice-Hall, Englewood Cliffs, NJ, 1991, Appendix A.

9. As an example, see Asato, M., Final Report. Spacecraft Design and Cost Model, USC, 1989.

Part two

New domains, new insights

Part 2 explores from an architectural point of view four domains beyond those of aerospace and electronics, the sources of most examples and writings to date. The chapters can be read for several purposes. For a reader familiar with a domain, there are broadly applicable heuristics for more effective architecting of its products. For ones unfamiliar with it, there are insights to be gained from understanding problems differing in the degree but not in kind from one's own. To coin a metaphor, if the domains can be seen as planets, then this part of the book corresponds to comparative planetology, the exploration of other worlds to benefit one's own. The chapters can be read for still another purpose, as a template for exploring other, equally instructive domains. An exercise for that purpose can be found at the end of Chapter 6, Software Systems.

From an educational point of view, this part is a recognition that one of the best ways of learning is by example, even if the example is in a different field or domain. One of the best ways of understanding another discipline is to be given examples of problems it solves. And one of the best ways of learning architecting is to recognize that there are architects in every domain and at every level from which others can learn and with whom all can work. At the most fundamental level, all speak the same language and carry out the same process, systems architecting. Only the examples are different.

Chapter 3 explores systems for which form is predetermined by a builder's perceptions of need. Such systems differ from those that are driven by client purposes by finding their end purpose only if they succeed in the marketplace. The uncertainty of end purpose has risks and consequences that are the responsibility of architects to help reduce or exploit. Central to doing so are the protection of critical system parameters and the formation of innovative architecting teams.

Chapter 4 highlights the fact that manufacturing has its own waterfall, quasi-independent of the more widely discussed product waterfall, and that these two waterfalls must intersect properly at the time of production. A spiral-to-circle model is suggested to help understand the integration of hardware and software. Ultraquality and feedback are shown to be the keys to both lean manufacturing and flexible manufacturing, with the latter likely to need a new information flow architecture if it is to succeed.

Chapter 5, Social Systems, introduces a number of new insights to those of the more technical domains. Economic questions and value judgments play a much stronger role here, even to the point of outright veto of otherwise worthwhile systems. A new tension comes to center stage, one central to social systems but too often downplayed in others until too late — the tension between facts and perceptions. It is so powerful in defining success that it can virtually mandate system design and performance, solely because of how that architecture is perceived.

Chapter 6, Software Systems, serves to introduce this rapidly expanding domain as it increasingly becomes the center of almost all modern systems designs. Consequently, whether they stand alone or as part of a larger system, software systems must accommodate to continually changing technologies and product usage. In very few other domains is annual, much less monthly, wholesale replacement of a deployed system economically feasible or even considered. In point of fact it is considered normal in software systems, precisely because of software's unique ability to continuously and rapidly evolve in response to changes in technology and user demands. Software has another special property: it can be as hard or as soft as needed. It can be hard wired if certification must be precise and unchanging. Or it can be as soft as a virtual environment molded at the will of a user. For these and other reasons, software practice is heavily dependent on heuristic guidelines and organized, layered modeling. It is a domain in which architecting development is very active, particularly in progressive modeling and rapid prototyping. The nature of modern software practice and its central role in new complex systems make Chapter 6 a natural lead into Part 3, Models and Representations.

chapter three

Builder-architected systems

No system can survive that doesn't serve a useful purpose.

Harry Hillaker*

Introduction: the form-first paradigm

The classical architecting paradigm is not the only way to create and build large complex systems, nor is it the only regime in which architects and architecting are important. A different architectural approach, the "form first", begins with a builder-conceived architecture in mind, rather than with a set of client-accepted purposes. Its architects are generally members of the technical staff of the company. Their client is the company itself.

The desired outcome of form-first architecting is usually an evolution of an existing architecture, a natural inclination of successful builders. But in other important cases, the motivation comes from the belief of a builder that a new technology can make possible new system functions for new markets. Major examples are commercial aircraft, personal computers, smart automobiles, and the continually evolving software for each. Generally speaking, the preconceived architecture is maintained with few modifications until an essentially completed system is presented to potential buyers. Because buyers determine the real value of a system, end purpose and survival are determined last, not first.

The form-first, technology-driven approach clearly is of appreciable risk for the builder — the potential buyer can always reject the system, use it for quite different purposes than intended, or force its price down to match its buyer-perceived value. It is this element of high risk that makes systems architecting so essential. The greater the risk, the greater the need to maintain the integrity of the envisioned

---

* Chief architect, McDonnell Douglas F-16 Fighter. As stated in a USC Systems Architecting lecture, November 1989.

system to reduce that risk. The greater is the need for well-defined and prioritized system-level functions and behavior.

The tools and methods to do so, as in the classical paradigm, are design management, heuristic reasoning, model building, certification of conformance to the initial vision of the builder, and controlled change to new purposes. In addition, form-first architecting will require early identification and proprietary control of critical architectural features.

Form-first systems

As indicated earlier, most form-first, builder-proposed systems are evolutions of an existing architecture or product line to enhance existing markets. To a large extent, the customers and purposes are understood, improvements are incremental, and risks are moderate. Familiar system-level functions, performance, quality, price, and availability are almost always retained if not improved. Extensive reuse of existing modules is common and cost effective because interfaces are largely unchanged. Incremental value generally matches incremental price. The architect, as always close to the customer but in this case employed by the builder, safeguards the product line against competition and circumstance and helps assure system integrity during development. The architecting process generally follows the classical paradigm as part of a customer-reactive, but product-defensive, strategy.

The next level of architecting intensity is reached when the builder's motivation is to use or modify an existing product line to reach uncertain or "latent" markets in which the potential customer must see the end product before judging its value. To succeed in the new venture, architecting must be proactive, making suggestions or proposing options without seriously violating the constraints of an existing product line. Hewlett Packard has developed this architecting technique in a novel way. Within a given product line, say, of an analytic instrument, a small set of feasible "reference" architectures is created, one of which should be close to satisfying a set of needs of a specific customer. Small changes in that architecture then enable tailoring to customer-expressed priorities. Another advantage: latent markets discovered in the process can be exploited by expansion of baseline production.

Of greatest risk are those form-first, technology-driven systems that create major qualitative changes in system-level behavior, changes in kind rather than of degree. Systems of this type almost invariably require across-the-board new starts in design, development, and use. They most often arise when radically new technologies become available — such as jet engines, new materials, microprocessors, lasers, software architectures, and intelligent machines. Although new technologies are infamous for creating unpleasant technological and even sociological surprises, by far the greatest single risk in these systems is one of *timing*. Even if the form is feasible, introducing a new product either too early or too late can be punishing. Douglas Aircraft Company was too late into jet aircraft, losing out for years to The Boeing Company. Innumerable small companies have been too early, unable to sustain themselves while waiting for the technologies to evolve into engineered products. High-tech defense systems, most often due to a premature commitment to a critical new technology, have suffered serious cost overruns and delays.

The second greatest single risk is in not recognizing that before they are completed, technology-driven architectures will require much more than just replacing, one at a time, components of an older technology for those of a newer one. Painful experience shows that without widespread changes in the system and its management, technology-driven initiatives seldom meet expectations and too often cost more for less value. As examples, direct replacements of factory workers with machines,[1] of vacuum tubes with transistors, of large inventories with Just-In-Time deliveries, and of experienced analysts with computerized management information systems all collapsed when attempted by themselves in a system that was otherwise unchanged. They succeeded only when incorporated in concert with other matched and planned changes. It is not much of an exaggeration to say that the successes were well architected; the former failures were not.

In automobiles, the most recent and continuing change is the insertion of ultraquality electronics and software between the driver and the mechanical subsystems of the car. This remarkably rapid evolution removes the driver almost completely from contact with, or direct physical control of, those subsystems. It considerably changes such overall system characteristics as fuel consumption, aerodynamic styling, driving performance, safety, and servicing and repair — as well as the design of such possibly unexpected elements as engines, transmissions, tires, dashboards, seats, passenger restraints, and freeway exits. As a point of fact, the automotive industry expects that by the turn of the century more than 93% of all automotive equipment will be computer controlled,[2] a trend evidently welcomed and used by the general public or it would not be done. A telling indicator of the public's perception of automotive performance and safety is the recent, virtually undisputed increase in national speed limits. Safe, long distance, highway travel at 70 mph (117 km/h) was rare, even dangerous, a decade ago. Even if the highways were designed for it, conventional cars and trucks were not. It is now common, safe, and legal. Perhaps the most remarkable fact about this rapid evolution is that most customers were never aware of it. This result came from a commitment to quality so high that a much more complex system could be offered that, contrary to the usual experience, worked far better than its simpler predecessor.

In aircraft, an equivalent, equally rapid, technolgy-driven evolution is "fly by wire", a change that, among other things, is forcing a social revolution in the role of the pilot and in methods of air traffic control. More is involved than the form-fit-function replacement of mechanical devices with a combination of electrical, hydraulic, and pneumatic units. Aerodynamically stable aircraft, which maintain steady flight with nearly all controls inoperative, are steadily being replaced with ones which are less stable, more maneuverable, and computer controlled in all but emergency conditions. The gain is more efficient, potentially safer flight. But the transition has been as difficult as that between visual and instrument-controlled flight.

In inventory control, a remarkable innovation has been the very profitable combination in one system of point-of-sale terminals, of a shift of inventory to central warehouses, and of Just-In-Time deliveries to the buyer. Note the word *combination*. None of the components has been particularly successful by itself. The risk here is greater susceptibility to interruption of supply or transportation during crises.

In communications, satellites, packet switching, and e-mail have combined for easier access to a global community, but with increasing concerns about privacy and security.

In all of these examples, far more is affected than product internals. Affected also are such externals as manufacturing management, equity financing, government regulations, and the minimization of environmental impact — to name but a few. These externals alone could explain the growing interest by innovative builders in the tools and techniques of systems architecting. How else to create well-balanced, well-integrated, financially successful, and socially acceptable total systems?

Consequences of uncertainty of end purpose

Uncertainty of end purpose, no matter what the reason, can have serious consequences. The most serious is the likelihood of serious error in decisions affecting system design, development, and production. Builder-architected systems are often solutions looking for a problem and hence are particularly vulnerable to the infamous "error of the third kind": working on the wrong problem.

Uncertainty in system purposes also weakens them as criteria for design management. Unless a well-understood basis for configuration control exists and can be enforced, system architectures can be forced off course by accommodations to crises of the moment. Some of the most expensive cases of record have been in attempts to computerize management information systems. Lacking clear statements of business purposes and market priorities, irreversible ad hoc decisions were made which so affected their performance, cost, and schedule that the systems were scrapped. Arguably, the best prevention against "system drift" is to decide on provisional or baseline purposes and stick to them. But what if those baseline purposes prove to be wrong in the marketplace?

Reducing the risks of uncertainty of end purpose

A powerful architecting guide to protect against the risk of uncertain purposes is the familiar **build in and maintain options** heuristic. With options available, early decisions can be modified or changed later. Other possibilities: build in options to stop at known points to guarantee at least partial satisfaction of user purposes without serious losses in time and money, for example, in data bases for accounting and personnel administration. Create architectural options that permit later additions, a favorite strategy for automobiles and trucks. Provisions for doing so are hooks in software to add applications and peripherals, scars in aircraft to add range and seats, shunts in electrical systems to isolate troubled sections, contingency plans in tours to accommodate cancellations, and forgiving exits from highways to minimize accidents.

In software, a general strategy is **Use open architectures. You will need them once the market starts to respond.** As will be seen, a further refinement of this domain-specific heuristic will be needed, but this simpler version makes the point for now.

And then there is the always-welcome heuristic: every once in a while, **Pause and reflect**. Reexamine the cost effectiveness of system features such as high precision pointing for weather satellites or cross-talk levels for tactical communication satellites.* Review why interfaces were placed where they were. Check for unstated assumptions such as the cold war continuing indefinitely** or the '60s generation turning conservative as it grew older.

Risk management by intermediate goals

Another strategy to reduce risk in the development of system-critical technologies is by scheduling a series of intermediate goals to be reached by precursor or partial configurations. For example, build simulators or prototypes to tie together and synchronize otherwise disparate research efforts.[3] Build partial systems, demonstrators, or models to help assess the sensitivity of customer acceptance to the builder's or architect's value judgments,[4] a widely used market research technique. And, as will be seen in Chapter 4, if these goals result in stable intermediate forms, they can be powerful tools for integrating hardware and software.

Clearly, precursor systems have to be almost as well architected as the final product. On the one hand, their failure in front of prospective customers can jeopardize future acceptance and ruin any market research program. As one heuristic derived from military

---

* In real life, both features proved to be unnecessary but could not be eliminated by the time that truth was discovered.
** A half-joking question in defense plannning circles used to be "What if peace broke out?" No more.

programs warns, **"The probability of an untimely failure increases with the weight of brass in the vincinity."** If precursors and demonstrators are to work well "in public", they better be well designed and well built.

On the other hand, a successful intermediate test can generate excessive confidence if a requirement yet to be tested is the most critical. In one case, a satelllite control system successfully and very publicly demonstrated an ability to manage one satellite at a time; the critical task, however, was to control multiple, different satelliltes, which it subsequently flunked. Massive changes in the system as a whole were required. In another similar case, a small launch vehicle, successful as a demonstrator, could not be scaled up to full size nor full capability for embarassingly basic mechanical and materials reasons.

These kinds of experiences led to the admonition: **Do the hard parts first,** an extraordinarily difficult heuristic to satisfy if the hard part is a unique function of the system as a whole. Such has been the case for a near-impenetrable missile defense system, stealthy aircraft, a general aviation air traffic control system, a computer operating system, and a national tax reporting system. The only credible precursor, to demonstrate the hard parts, had to be almost as complete as the final product.

In risk management terms, if the hard parts are, perhaps necessarily, left to last, then the risk level remains high and uncertain to the very end. The justification for the system therefore must be very high and the support for it very strong or its completion will be unlikely. For private businesses this means high-risk venture capital. For governments it means support by the political process, a factor in system acquisition which few architects, engineers, and technical managers understand. Chapter 10, The Political Process and Systems Architecting, is a primer on the subject.

The "what next?" quandary

One of the most serious long-term risks faced by a builder of a successful, technology-driven system is the lack of, or failure to win a competition for, a successor or follow-on to the original success.

The first situation is well exemplified by a start-up company's lack of a successor to its first product caused by a lack of resources in its early, profitless years to support more than one research and development effort. Ironically, the more successful the initial product, the more competition it will attract from established and well-funded producers anxious to profit from a sure thing. Soon the first product will be a "commodity", something which many companies can produce at a rapidly decreasing cost and risk. Unable to repeat the first success, the start-up enterprise fails or is bought up at fire-sale prices when the innovator can no longer meet payroll. Common. Sad. Avoidable? Possibly.

The second situation is the all-too-frequent inability of a well-established company which had been successfully supplying a market-valued system to win contracts for its follow-on. In this instance, the very strength of the successful system, a fine architecture matched with an efficient organization to build it, can be its weakness in a time of changing technologies and market needs. The assumptions and constraints of the present architecture can become so ingrained in the thinking of participants that options simply don't surface.[5] Ironic but true, the architecture will have been too successful for the company's own good.

In both situations, the problem is architectural, as is its alleviation.

For the innovative company, it is a matter of control of critical architectural features. For the successful first producer, it is a matter of knowing, well ahead of time, when purposes have changed enough that major re-architecting, probably by a new team, may be required. Each situation will be considered in turn.

Controlling the critical features of the architecture

The critical part of the answer to the start-up company's "what next" quandary is control of the architecture of its product through proprietary ownership of its basic features.[6] Examples of such features are computer operating systems, interface characteristics, communication protocols, microchip configurations, proprietary materials, and unique and expensive manufacturing capabilities. Good products, while certainly necessary, are not sufficient. They must also arrive on the market as a steadily improving product line, one that establishes, de facto, an architectural standard.

Surprisingly, one way to achieve that objective is to use the competition instead of fighting it. Because success invites competition, it may well be better for a start-up company — or project within a company — to have profitable competition dependent, through licensing, upon a proprietary architecture than to have well-financed competition incentivized to seek architectural alternatives. Finding architectural alternatives takes time. But licensing encourages the competition to find new applications, add peripherals, and develop markets, further strengthening the architectural base, adding to the source company's profits and its own development base. Heuristically:

Successful architectures are proprietary, but open.*

This strategy was well exemplified by IBM in opening and licensing its personal computer (PC) operating system while Apple refused to do so for its Macintosh. The resultant widespread cloning of the PC expanded not only the market as a whole, but IBM's share of it. The Apple share

---

* "Open" here means adaptable, friendly to add-ons, and selectively expandable in capability.

dropped. A different kind of architectural control is exemplified by the Bell telephone system with its technology generated by the Bell Laboratories, its equipment produced largely by Western Electric, and its architectural standards maintained by usage and regulation. Others include Xerox in copiers, Kodak in cameras, Intel in microprocessors, Hewlett Packard in instruments, and so on. All these product line companies began small, controlled the basic features, and prospered.

Thus, for the innovator the essentials for continued success are not only a good product, but also the generation, recognition, and control of its basic architectural features. Without these essentials, there may never be a successor product. With them, many product architectures, as architecturally controlled product lines, have lasted for years following the initial success — which adds even more meaning to **There's nothing like being the first success.**[7]

Abandonment of an obsolete architecture

A different risk-reduction strategy is needed for the company which has established and successfully controlled a product line architecture[8] and its market, but is losing out to a successor architecture that is proving to be better in performance, cost, and/or schedule. There are many ways that this can happen. Perhaps the purposes that original architecture has satisfied can better be done in other ways. Typewriters have largely been replaced by personal computers. Perhaps the conceptual assumptions of the original architecture no longer hold. Energy may no longer be cheap. Perhaps competitors found a way of bypassing the original architectural controls with a different architecture. Personal computers destroyed the market for Wang word processors. The desktop metaphor for personal computers revolutionized their user friendliness and their market. And, as a final example, cost risk considerations precluded building larger and larger spacecraft for the exploration of the solar system.

To avoid being superceded architecturally requires a strategy, worked out well ahead of time, to set to one side or cannibalize that first architecture, *including the organization matched with it*, and to take preemptive action to create a new one. The key move is the well-timed establishment of an innovative architecting team, unhindered by past success and capable of creating a successful replacement. Just such a strategy was undertaken by Xerox in a remake of the corporation as it saw its copier architecture start to fade. It thereby redefined itself as "the document company".[9]

Creating innovative teams

Clearly the personalities of members of any team, particularly an innovative architecting team, must be compatible. A series of USC Research

Reports[10] by Jonathan Losk, Tom Pieronek, Kenneth Cureton, and Norman P. Geis, based on the Myers-Briggs Type Indicator (MBTI), strongly suggest that the preferred personality type for architecting team membership is NT.[11] That is, members should tend toward systematic and strategic analysis in solving problems. As Cureton summarizes,

> Systems architects are made and not born, but some people are more equal than others in terms of natural ability for the systems architecting process, and MBTI seems to be an effective measure of such natural ability. No single personality type appears to be the 'perfect' systems architect, but the INTP personality type often possesses many of the necessary skills.

Their work also shows the need for later including an ENTP, a "field marshall" or deputy project manager, not only to add some practicality to the philosophical bent of the INTPs, but to help the architecting team work smoothly with the teams responsible for building the system.

Creating *innovative* teams is not easy, even if the members work well together. The start-up company, having little choice, depends on good fortune in its recruiting of charter members. The established company, to put it bluntly, has to be willing to change how it is organized and staffed from the top down, based almost solely on the conclusions of a presumably innovative team of "outsiders", albeit individuals chartered to be such. The charter is a critical element, not so much in defining new directions as in defining freedoms, rights of access, constraints, responsibilities, and prerogatives for the team. For example, can the team go outside the company for ideas, membership, and such options as corporate acquisition? To whom does the team respond and report — and to whom does it not? Obviously, the architecting team better be well designed and managed. Remember, if the team does not succeed in presenting a new and accepted architecture, the company may well fail.

One of the more arguable statements about architecting is the one by Frederick P. Brooks, Jr. and Robert Spinrad that "the best architectures are the product of a single mind." For modest-sized projects that statement is reasonable enough. But not for larger ones. The complexity and work load of creating large, multidisciplinary, technology-driven architectures would overwhelm any individual. The observation of a single mind is most easily accommodated by a simple but subtle change from "a single mind" to "a team of a single mind." Some would say "of a single vision" composed of ideas, purposes, concepts, presumptions, and priorities.

In the simplest case, the single vision would be that of the chief architect and the team would work to it. For practical as well as team cohesiveness reasons, the single vision needs to be a shared one. In no system is that more important than in the entrepreneurially motivated one. There will always be a tension between the more thoughtful architect and the more action-oriented entrepreneur. Fortunately, achieving balance and compromise of their natural inclinations works in the system's favor.

An important corollary of the shared vision is that the architecting team itself, and not just the chief architect, must be seen as creative, communicative, respected, and of a single mind about the system-to-be. Only then can the team be credible in fulfilling its responsibilities to the entrepreneur, the builder, the system, and its many stakeholders. Internal power struggles, basic disagreements on system purpose and values, and advocacies of special interests can only be damaging to that credibility.

As Ben Bauermeister, Harry Hillaker, Archie Mills, Bob Spinrad,[12] and other friends have stressed in conversations with the author, innovative teams need to be cultural in form, diverse in nature, and almost obsessive in dedication.

By cultural is meant a team characterized by informal creativity, easy interpersonal relationships, trust, and respect, all characteristics necessary for team efficiency, exchange of ideas, and by personal identification with a shared vision. To identify with a vision, they must deeply believe in it and in their chief. The members either acknowledge and follow the lead of their chief or the team disintegrates.

Diversity in specialization is to be expected; it is one of the reasons for forming a team. Equally important, a balanced diversity of style and programmatic experience is necessary to assure openmindedness, to spark creative thinking in others, and to enliven personal interrelationships. It is necessary, too, to avoid the "group think" of nearly identical members with the same background, interests, personal style, and devotion to past architectures and programs. Indeed, team diversity is one of the better protections against the second-product risks mentioned earlier.

Consequently, an increasingly accepted guideline is that to be truly innovative and competitive in today's world: **The team that created and built a presently successful product is often the best one for its evolution — but seldom for creating its replacement.**

A major challenge for the architect, whether as an individual or as the leader of a small architecting team, is to maintain dedication and momentum not only within the team, but also within the managerial structure essential for its support. The vision will need to be continually restated as new participants and stakeholders arrive on the scene — engineers, managers active and displaced, producers, users, even new clients. Even more difficult, it will have to be transformed as the system proceeds from a dream to a concrete entity, to

a profit maker, and finally to a quality production. Cultural collegi-
ality will have to give way to the primacy of the bottom line and
finally to the necessarily bureaucratic discipline of production. Yet
the integrity of the vision must never be lost or the system will die.

Systems architecting and basic research

One other relationship should be established, that between architects
and those engaged in basic research and technology development. Each
group can further the interests of the other.

New technologies enable new architectures, though not singly nor
by themselves. Consider solid-state electronics, fiber optics, software
languages, and molecular resonance imaging for starters. And innovative architectures provide the rationale for underwriting research, often
at a very basic level. Yet, though both innovative architecting and basic
research explore the unknown and unprecedented, there seems to be
little early contact between their respective architects and researchers.
The architectures of intelligent machines, the chaotic aerodynamics of
active surfaces, the sociology of intelligent transportation systems, and
the resolution of conflict in multimedia networks are examples of presumably common interests. Universities might well provide a natural
meeting place for seminars, consulting, and the creation and exchange
of tools and techniques.

New architectures, driven by perceived purposes, sponsor more
basic research and technology development than is generally acknowledged. Indeed, support for targeted basic research undoubtedly
exceeds that motivated by scientific inquiry. Examples abound in communications systems which sponsor coding theory, weapons systems
which sponsor materials science and electromagnetics, aircraft which
sponsor fluid mechanics, and space systems which sponsor the fields
of knowledge acquisition and understanding.

It is therefore very much in the mutual interest of professionals in
R & D and systems architecting to know each other well. Architects
gain new options. Researchers gain well-motivated support. Enough
said.

Heuristics for technology-driven systems

General

- An insight is worth a thousand market surveys.
- Success is defined by the customer, not by the architect.
- In architecting a new program, all the serious mistakes are made
  in the first day.
- The most dangerous assumptions are the unstated ones.

- The choice between products may well depend upon which set of drawbacks the users can handle best.
- As time to delivery decreases, the threat to user utility increases.
- If you think your design is perfect, it's only because you haven't shown it to someone else.
- If you don't understand the existing system, you can't be sure you are building a better one.
- Do the hard parts first.
- Watch out for domain-specific systems. They may become traps instead of useful system niches, especially in an era of rapidly developing technology.
- The team that created and built a presently successful product is often the best one for its evolution — but seldom for creating its replacement. (It may be locked into unstated assumptions that no longer hold.)

Specialized

From Morris and Ferguson 1993:

- Good products are not enough. (Their features need to be owned.)
- Implementations matter. (They help establish architectural control.)
- Successful architectures are proprietary, but open. (Maintain control over the key standards, protocols, etc., that characterize them but make them available to others who can expand the market to everone's gain.)

From Chapters 2 and 3:

- Use open architectures. You will need them once the market starts to respond.

Summary

Technology-driven, builder-architected systems, with their greater uncertainty of customer acceptance, encounter greater architectural risks than those that are purpose-driven. Risks can be reduced by the careful inclusion of options, the structuring of their innovative teams, and the application of heuristics found useful elsewhere. At the same time, they have lessons to teach in the control of critical system features and the response to competition enabled by new technologies.

Exercises

1. The architect can have one of three relationships to the builder and client. The architect can be a third party, can be the builder, or can be the client. What are the advantages and disadvantages of each relationship? For what types of system is one of the three relationships necessary?

2. In a system familiar to you, discuss how the architecture can allow for options to respond to changes in client demands. Discuss product vs. product-line architecture as a response to the need for options. Find examples among systems familiar to you.

3. Architects must be employed by builders in commercially marketed systems because customers are generally unwilling to sponsor development; they purchase systems after seeing the finished product. But placing the architect in the builder's organization will tend to dilute the independence needed by the architect. What organizational approaches can help to maintain independence while also meeting the needs of the builder organization?

4. The most difficult type of technology-driven system is one that does not address any existing market. Examine the history of both successful and failed systems of this type. Can any lessons be extracted?

Notes and References

1. Majchrzak, A., *The Human Side of Automation*, Jossey-Bass Publishers, San Francisco, 1988, 95. The challenge: 50–75% of the attempts at introducing advanced manufacturing technology into U.S. factories are unsuccessful, primarily due to lack of human resource planning and management.

2. Automobile Club of Southern California speaker, Los Angeles, August 1995.

3. The Goldstone California planetary radar system of the early 1960s tied together antenna, transmitter, receiver, signal coding, navigation, and communications research and development programs. All had to be completed and integrated into a working radar at the fixed dates when target planets were closest to the Earth and well in advance of commitment for system support of communication and navigation for as-yet-not-designed spacecraft to be exploring the solar system. Similar research coordination can be found in NASA aircraft, applications software, and other rapid prototyping efforts.

4. Hewlett-Packard marketeers have long tested customer acceptance of proposed instrument functions by presenting just the front panel and asking for comment. Both parties understood that what was behind the panel had yet to be developed, but could be. A more recent adaptation of that technique is the presentation of "reference architectures".

5. An example can be found in a series of satellite survelliance and remote sensing programs in which the government elected to compete follow-ons to highly successful initial systems. A series of eight initial contractors, all highly qualified, lost the eight follow-on contracts. The winning contractors evidently rejected bidding on an evolutionary change, sure to be won by the initial contractor, and bid and won on highly innovative architectures exhibiting superior performance, cost, and schedule.

6. Morris, C. R. and Ferguson, C. H., How architecture wins technology wars, *Harvard Business Review,* March–April 1993; and its notably supportive follow-up review, Will architecture win the technology wars?, *Harvard Business Review,* May–June 1993. A seminal article on the subject.

7. Rechtin, E., *Systems Architecting, Creating & Building Complex Systems,* Prentice-Hall, Englewood Cliffs, NJ, 1991, 301.

8. It is important to distinguish between a product line architecture and a product in that line. It is the first of these that is the subject here. Hewlett Packard is justly famous for the continuing creation of innovative products to the point where half or less of their products on the market were there 5 years earlier. Their product *lines* and their architectures, however, are much longer lived. Even so, these, too, were replaced in due course, the most famous of which was the replacement of "dumb" instruments with smart ones.

9. Described by Robert Spinrad in a 1992 lecture to a USC systems architecting class. Along with Hewlett Packard, Xerox demonstrates that it is possible to make major architectural, or strategic, change without completely dismantling the organization. But the timing and nature of the change is critical. Again, **a product architecture and the organization that produces it must match, not only as they are created but as they decline**. For either of the two to be too far ahead or behind the other can be fatal.

10. Geis, N. P., Profiles of Systems Architecing Teams, USC Research Report, June 1993, extends and summarizes the earlier work of Losk, Pieronek, and Cureton.

11. It is beyond the scope of this book to show how to select individuals for teams, only to mention that widely used tools are available for doing so. For further information on MBTI and the meaning of its terms see Myers, I., Briggs, K., and McCaulley, *Manual: A Guide to the Development and Use of the Myers-Briggs Type Indicator,* Consulting Psychologists Press, Palo Alto, CA, 1989.

12. Benjamin Bauermeister is an entrepreneur and architect in his own software company. He sees three successive motivations as product development proceeds: product, profit, and production corresponding roughly to cultural, bottom line, and bureaucratic as successive organization forms. Harry Hillaker is the retired architect of the Air Force F-16 fighter and a regular lecturer in systems architecting. Archie W. Mills is a middle-level manager at Rockwell International and the author of an unpublished study of the practices of several Japanese and American companies over the first decades of this century in aircraft and electronics systems. Bob Spinrad is a XEROX executive and former director of XEROX Palo Alto Research Center (PARC).

chapter four

Manufacturing systems

Introduction: the manufacturing domain

Although manufacturing is often treated as if it were but one step in the development of a product, it is also a major system in itself. It has its own architecture.[1] It has a system function that its elements cannot perform by themselves — making other things with machines. And it has an acquisition waterfall for its construction quite comparable to those of its products.

From an architectural point of view, manufacturing has long been a quiet field. Such changes as were required were largely a matter of continual, measurable, incremental improvement — a step at a time on a stable architectural base. Though companies came and went, it took decades to see a major change in its members. The percentage of sales devoted to research and advanced development for manufacturing, per se, was small. The need was to make the classical manufacturing architecture more effective, that is, to evolve and engineer it.

A decade or so ago the world that manufacturing had supported for almost a century changed — and on a global scale. Driven by new technologies in global communications, transportation, sources, markets, and finance, global manufacturing became practical and then, shortly thereafter, dominant. It quickly became clear that qualitative changes were required in manufacturing architectures if global competition were to be met. In the order of appearance, the architecural innovations were ultraquality,[2] dynamic manufacturing,[3] lean production,[4] and "flexible manufacturing".* The results to date, demonstrated first by the Japanese, have been greatly increased profits and market share, and sharply decreased inventory and time-to-market. Each of these innovations will be presented in turn.

Even so, rapid change is still underway. As seen on the manufacturing floor, manufacturing research has yet to have much of an effect. Real-time software is still a rarity. Trend instrumentation, self-diagnosis, and self-correction, particularly for ultraquality systems, are far

---

* Producing different products on demand on the same manufacturing line. See "Flexible Manufacturing" section of this chapter.

from comonplace. So far, the most sensitive tool is the failure of the product being manufactured.

Architectural innovations in manufacturing

Ultraquality systems

At the risk of oversimplification, a common perception of quality is that quality costs money; that is, that quality is a trade-off against cost and/or profit. Not coincidently, there is an accounting category called "cost of quality". A telling illustration of this perception is the "DeSoto Story". As the story goes, a young engineer at a DeSoto automobile manufacturing plant went to his boss with a bright idea on how to make the DeSoto a more reliable automobile. The boss's reply: "Forget it, kid. If it were more reliable it would last more years and we would sell fewer of them. It's called planned obsolescence." DeSoto no longer is in business but the perception remains in the minds of many manufacturers of many products.

The difficulty with this perception is partly traceable to the two different aspects of quality. The first is quality associated with features like leather seats and air conditioning. Yes, those features cost money but the buyer perceives them as value added and the seller almost always *makes* money on them. The other aspect of qualilty is absence of defects. As it has now been shown, absence of defects *also* makes money for both seller and buyer, through reductions in inventory, warranty costs, repairs, documentation, testing, and time to market — *provided that* the level of product quality is high enough and that the whole development and production process is architected at that high level.

To understand why absence of defects makes money, imagine a faultless process that produces a product with defects so rare that it is impractical to measure them; that is, none are anticipated within the lifetime of the product. Testing can be reduced to the minimum required to certify system-level performance of the first unit. Delays and their costs can be limited to those encountered during development; if and when later defects occur, they can be promptly diagnosed and permanently fixed. "Spares" inventory, detailed parts histories, and statistical quality control can be almost nonexistent. First-pass manufacturing yield can be almost 100% instead of today's highly disruptive 20–70%. Service in the field is little more than replacing failed units, free.

The best possible measurement of ultraquality is then an end system-level test of the product itself. Attempting to measure defects at any subsystem level would be a waste of time and money — defects would be too rare. Redundancy and fault tolerant designs would be unnecessary. Indeed, they would be impractical because, without an expectation of a specific failure (which then should be fixed), protection against rare and unspecified defects is not cost effective.

To some readers, this ultraquality level may appear to be hopelessly unrealistic. Suffice it to say that it has been approached for all practical purposes. For decades, no properly built unmanned satellite or spacecraft failed because of a defect known before launch (they would have been fixed beforehand). Microprocessors with millions of active elements, sold in the millions, now outlast the computers for which they were built. Like the satellites, they become technically and financially obsolete before they fail. Television sets are produced with a production line yield of over 99%, far better than the 50% yield of a decade ago, with a major improvement in cost, productivity, and profit.

Today's challenge, then, is to achieve and maintain such ultraquality levels even as systems become more complex. Techniques have been developed which certainly help.* More recently, two more techniques have been demonstrated that are particularly applicable to manufacturing.

The first is **Everyone in the production line is both a customer and a supplier,** a customer for the preceding worker and a supplier for the next. Its effect is to place quality assurance where it is most needed, at the source.

The second is **The Five Why's,** a diagnostic procedure for finding the basic cause of a defect or discrepancy. Why did this occur? Then why did that, in turn, occur? Then, why *that?* and so on until the offending causes are discovered and eliminated.

To these techniques can be added a relatively new understanding: **Some of the worst failures are system failures,** that is, they come from the interaction of subsystem deficiencies which of themselves do not produce an end system failure, but together can and do. Four catastrophic civil space system failures were of this kind: Apollo 1, Apollo 13, Challenger, and the Hubble Telescope. For Tom Clancy buffs, just such a failure almost caused World War III in his *Debt of Honor.* In all these cases, had any one of the deficiencies not occurred, the near-catastrophic end result could not have occurred. That is, though each deficiency was necessary, none were sufficient for end failure. As an admonition to future failure review boards, until a diagnosis is made that indicates that the set of presumed causes is both necessary and

---

* Rechtin, E., 1991 Chapter 8, p. 160–187. One technique mentioned there — fault tolerance through redundancy — has proved to be less desirable than proponents had hoped. Because fault tolerant designs "hide" single faults by working around them, they accumulate until an overall system failure occurs. Diagnostics then become very much more difficult. Symptoms are intertwined. Certification of complete repair cannot be guaranteed because successful-but-partial operation again hides undetected (tolerated) faults. The problem is most evident in servicing modern, microprocessor-rich, automobile controls. The heuristic still holds, **Fault avoidance is preferable to fault tolerance in system design.**

sufficient — and that no other such set exists — the discovery-and-fix process is incomplete and ultraquality is not assured.

Succcessful ultraquality has indeed been achieved, but there is a price that must be paid. Should ultraquality *not* be produced at any point in the whole production process, the process may collapse. Therefore, when something does go wrong, it must be fixed immediately; there are no cushions of inventory, built-in holds, full-time expertise, or planned work-arounds. Because strikes and boycotts can have instantaneous effects, employee, customer, and management understanding and satisfaction are essential. Pride in work and dedication to a common cause can be of special advantage, as has been seen in the accomplishments of the zero-defect programs of World War II, the American Apollo lunar landing program, and the Japanese drive to world-class products.

In a sense, ultraquality-built systems are fine tuned to near perfection with all the risks thereof. Just how much of a cushion or insurance policy is needed for a particular system is an important value judgment that the architect must obtain from the client, the earlier the better. That judgment has strong consequences in the architecture of the manufacturing system. Clearly, then, ultraquality architectures are very different from the statistical quality assurance architectures of only a few years ago.*

Most important for what follows, it is unlikely that either lean production or flexible manufacturing can be made competitive at much less than ultraquality levels.

Dynamic manufacturing systems

The second architectural change in manufacturing systems is from comparatively static configurations to dynamic, virtually real-time ones. Two basic architectural concepts now become much more important. The first concerns intersecting waterfalls; the second, feedback systems.

Intersecting waterfalls

The development of a manufacturing system can be represented as a separate waterfall, distinct from, and intersecting with, that of the product it makes. Figure 4.1 depicts the two waterfalls, the process (manufacturing) diagonally and the product vertically, intersecting at the time and point of production. The manufacturing one is typically longer in time, often decades, and contains more steps than the rela-

---

* One of the authors, a veteran of the space business, recently visited two different manufacturing plants and correctly predicted the plant yields (acceptance vs. start rates) by simply looking at the floors and at the titles (not content) and locations of a few performance charts on the walls. In the ultraquality case, the floors were painted white; the charts featured days since last defect instead of running-average defect rates; and the charts were placed at the exits of each work area. Details, but indicative.

tively shorter product sequence (months to years) and may end quite differently (the plant is shut down and demolished). Sketching the product development and the manufacturing process as two intersecting waterfalls helps emphasize the fact that manufacturing has its own steps, time scales, needs, and priorities distinct from those of the product waterfall. It also implies the problems its systems architect will face in maintaining system integrity, in committing well ahead to manufacture products not yet designed, and in adjusting to comparatively abrupt changes in product mix and type. A notably serious problem is managing the introduction of new technologies, safely and profitably, into an inherently high-inertia operation.

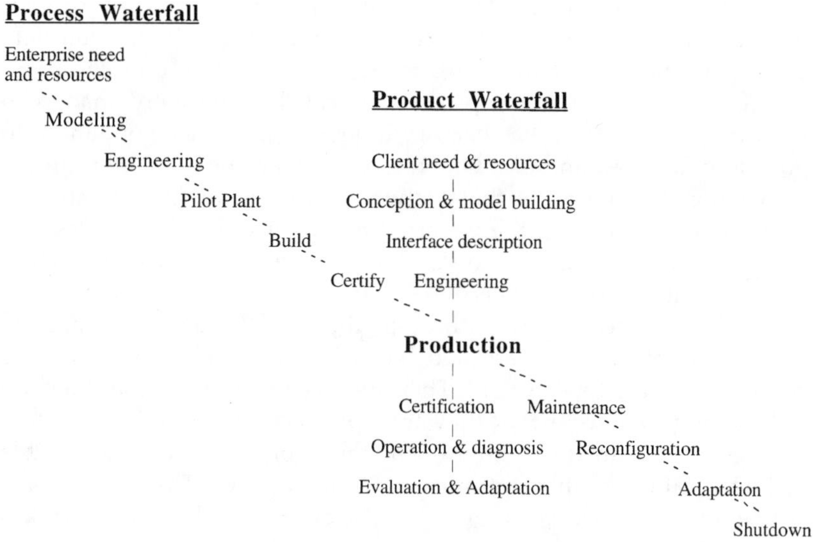

Process Waterfall

Enterprise need
and resources

Product Waterfall

    Modeling

        Engineering   Client need & resources

           Pilot Plant   Conception & model building

               Build   Interface description

           Certify   Engineering

Production

        Certification   Maintenance

      Operation & diagnosis   Reconfiguration

    Evaluation & Adaptation   Adaptation

                                Shutdown

Figure 4.1   The intersecting process and product waterfalls.

There are other differences. Manufacturing certification must begin well before product certification or the former will preclude the latter; in any case, the two must interact. The product equivalent of plant demolition, not shown in the figure, is recycling, both now matters of national law in Europe. Like certification, demolition is important to plan early, given its collateral human costs in the manufacturing sector. The effects of environmental regulations, labor contracts, redistribution of usable resources, retraining, right-sizing of management, and continuing support of the customers are only a few of the manufacturing issues to be resolved — and well before the profits are exhausted.

Theoretically if not explicitly, these intersecting waterfalls have existed since the beginning of mass production. But not until recently have they been perceived as having equal status, particularly in the U.S. Belatedly, that perception is changing, driven in large part by the establishment of global manufacturing — clearly not the same system as a wholly owned shop in one's own backyard. The change is magnified by the widespread use of sophisticated software in manufacturing, a boon in managing inventory, but a costly burden in reactive process control.[5] Predictably, software for manufacturing process and control, more than any element of manufacturing, will determine the practicality of flexible manufacturing. As a point in proof, Hayes, Wheelright, and Clark[6] point out that a change in the architecture of [even] a mass production plant, particularly in the software for process control, can make dramatic changes in the capabilities of the plant without changing either the machinery or layout.

The development of manufacturing system software adds yet another production track. The natural development process for software generally follows a spiral,* certainly not a conventional waterfall, cycling over and over through functions, form, code (building), and test. Figure 4.2's software spiral is typical. It is partially shaded to indicate that one cycle has been completed with at least one more to go before final test and use. One reason for such a departure from the conventional waterfall is that software, as such, requires almost no manufacturing, making the waterfall model of little use as a descriptor.** The new challenge is to synchronize the stepped waterfalls and the repeating spiral processes of software-intensive systems. One of the most efficient techniques is through the use of stable intermediate forms,[7] combining software and hardware into partial but working precursors to the final system. Their important feature is that they are *stable* configurations; that is, they are reproduceable, well-documented, progressively refined baselines — in other words, they are identifiable architectural waypoints and must be treated as such. They can also act as operationally useful configurations,*** built-in "holds" allowing lagging elements to catch up, or parts of strategies for risk reduction as suggested in Chapter 3.

The spiral-to-circle model

Visualizing the synchronization technique for the intersecting waterfalls and spirals of Figure 4.2 can be made simpler by modifying the spiral so that it remains from time to time in stable concentric circles on the four-quadrant process diagram. Figure 4.3 shows a typical

---

* See also Chapter 6 and Figure 6.1.
** Efforts to represent the manufacturing and product processes as spirals have been comparably unsuccessful. Given the need to order long-lead-time items, to "cut metal" at some point, and to write off the cost of multiple rapid prototypes, the waterfall is the depiction of choice.
*** Many Commanders of the Air Force Space and Missiles Division have insisted that all prototypes and interim configurations have at least some operational utility, if only to help increase the acceptance in the field once the final configuration is delivered. In practice, field tests of interim configurations almost always clarify, if not reorder, prior value judgments of what is most important to the end user in what the system can offer.

Process Waterfall

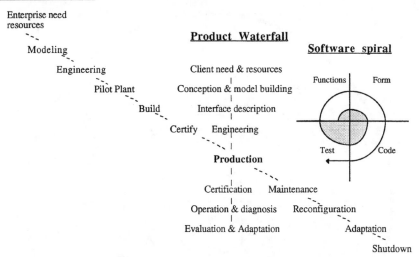

Figure 4.2 Product, process, and software system tracks.

development from a starting point in the "function" quadrant cycling through all quadrants three times — typical of the conceptualization phase — to the first intermediate form. There the development may stay for a while, changing only slightly, until new functions call for the second form, say an operational prototype. In Air Force procurement, that might be a "fly-before-buy" form. In space systems, it might be a budget-constrained "operational prototype" which is actually flown. In one program, it turned out to be the *only* system flown. But, to continue, the final form is gained, in this illustration, by way of a complete, four-quadrant cycle.

The spiral-to-circle model can show other histories; for example, a failure to spiral to the next form, with a retreat to the preceding one, possibly with less ambitious goals — or a transition to a still greater circle in a continuing evolution — or an abandonment of these particular forms with a restart near the origin.

Synchronization can also be helped by recognizing that cycling also goes on in the multistep waterfall model, except that it is depicted as feedback from one phase to one or more preceding ones. It would be quite equivalent to software quadrant spiraling if all waterfall feedback returned to the beginning of the waterfall, that is, to the system's initial purposes and functions, and from there down the waterfall again.

The circle-to-spiral model for software-intensive systems in effect contains both the expanding-function concept of software and the stepwise character of the waterfall. It also helps understand what and when hardware and software functions are needed in order to satisfy requirements by the *other* part of the system. For example, a stable intermediate

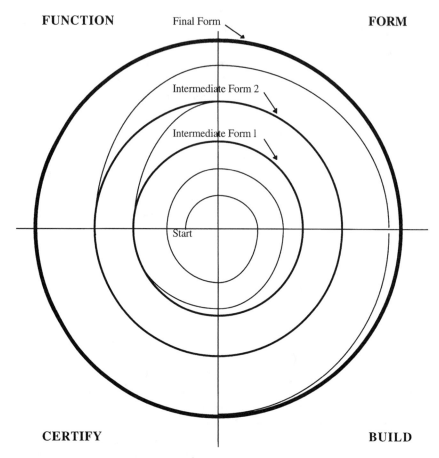

FUNCTION   Final Form   **FORM**

Intermediate Form 2

Intermediate Form 1

Start

CERTIFY | **BUILD**

Figure 4.3   The spiral-to-circle model.

software form of software should arrive when a stable, working form
of hardware arrives that needs that software, and vice versa.

It is important to recognize that this model, with its cross-project
synchronization requirement, is notably different from models of pro-
curements in which hardware and software developments can be relatively independent of each other. In the spiral-to-circle model, the
intermediate forms, both software and hardware, must be relatively
unchanging and bug free. A software routine that never quite settles
down or that depends upon the user to find its flaws is a threat, not a
help, in software-intensive systems procurement. A hardware element
that is intended to be replaced with a "better and faster" one later is
hardly better. Too many subsequent decisions may unknowingly rest
on what may turn out to be anomalous or misunderstood behavior of
such elements in system test.

To close this section, although this model may be relatively new,
the process that it describes is not. Stable intermediate forms, blocks

(I, II, etc.), or "test articles", as they are called, are built into many system contracts and perform as intended. Yet there remains a serious gap between most hardware and software developers in their understanding of each other and their joint venture. As the expression goes, "These guys just don't talk to each other." The modified spiral model should help both partners bridge that gap, to accept the reasons both for cycling and for steps, and to recognize that neither acquisition process can succeed without the success of the other.

There should be no illusion that the new challenge will be easy to meet. Intermediate software forms will have to enable hardware phases at specified milestones — not just satisfy separate software engineering needs. The forms must be stable, capable of holding at that point indefinitely, and practical as a stopping point in the acquisition process if necessary. Intermediate hardware architectures must have sufficient flexibility to accommodate changes in the software — as well as in the hardware. And finally, the architects and managers will have a continuing challenge in resynchronizing the several processes so that they neither fall behind nor get too far ahead.Well-architected intermediate stable forms and milestones will be essential.

Concurrent engineering

To return to Figure 4.2, this intersecting waterfall model also helps identify the source of some of the inherent problems in concurrent (simultaneous, parallel) engineering — in which product designers and manufacturing engineers work together to create a well-built product. Concurrent engineering in practice, certainly, has proven to be more than modifying designs for manufacturability. However defined, it is confronted with a fundamental problem, evident from Figure 4.2; namely, coordinating the two intersecting waterfalls and the spirals, each with different time scales, priorities, hardware, software, and profit-and-loss criteria. Because each track is relatively independent of the others, the incentives for each are to optimize locally even if in doing so results in an impact on another track or on the end product. After all, it is a human and organizational objective to solve one's own problems, to have authority reasonably commensurate with responsibilities, and to be successful in one's own right. Unfortunately, this objective forces even minor technical disagreements to higher, enterprise management where other considerations than just system performance come into play.

A typical example: A communications spacecraft design was proceeding concurrently in engineering and manufacturing until the question came up of the spacecraft antenna size. The communications engineering department believed that a 14-ft diameter was needed; the manufacturing department insisted that 10 ft was the practical limit. The difference in system performance was a factor of two in communications capability and revenue. The reason for the limit, it turned out, was that the manufacturing department had a first-rate subcontractor with all the equipment needed to build an excellent antenna, but no larger than 10 ft. To go larger would cause a measurable manufacturing cost overrun. The manufacturing manager was adamant about staying within his budget, having taken severe criticism for an overrun in the previous project. In any case, the added communications revenue gain was far larger than the cost of reequipping the subcontractor. Lacking a systems architect, the departments had little choice but to escalate the argument to a higher level where the larger antenna was eventually chosen and the manufacturing budget increased slightly. The design proceeded normally until software engineers wanted to add more memory well after manufacturing had invested heavily in the original computer hardware design. The argument escalated, valuable time was lost, department prerogatives were again at stake ....and so it went.

The example is not uncommon. A useful management improvement would have been to set up a trusted, architect-led team to keep balancing the system as a whole within broad top managment guidelines of cost, performance, risk, and schedule.

If so established, the architectural team's membership should include a corporate-level (or "enterprise") architect, the product architect, the manufacturing architect, and a few specialists in system-critical elements — and no more.[8] Such a structure does exist implicitly in some major companies, though seldom with the formal charter, role, and responsibilities of systems architecting.

Feedback systems

Manufacturers have long used feedback to better respond to change. Feedback from the customer has been, and is, used directly to maintain manufacturing quality and indirectly to accommodate changes in design. Comparably important are paths from sales to manufacturing and from manufacturing to engineering.

What is new is that the pace has changed. Multiyear is now yearly, yearly is now monthly, monthly is now daily, and daily — especially for ultraquality systems — has become hourly if not sooner. What was a temporary slowdown is now a serious delay. What used to affect only adjacent sectors can now affect the whole. What used to be the province of planners is now a matter of real-time operations.

Consequently, accustomed delays in making design changes, correcting supply problems, responding to customer complaints, introducing new products, reacting to competitors' actions, and the like were no longer acceptable. The partner to ultraquality in achieving global competitiveness was to counter the delays by anticipation of them —

in other words, to use anticipatory feedback in as close to real-time as practical. The most dramatic industrial example to date has been in Lean Production[9] in which feedback to suppliers, coupled with ultraquality techniques, cut the costs of inventory in half and resulted in across-the-board competitive advantage in virtually all business parameters. More recently, criteria for certification, or those of its predecessor, quality assurance, are routinely fed back to conceptual design and engineering — one more recognition that **quality must be designed in, not tested in.**

A key factor in the design of any real-time feedback system is loop delay, the time it takes for a change to affect the system "loop" as a whole. In a feedback system, delay is countered by anticipation based on anticipated events (like a failure) or on a trend derived from the integration of past information. The design of the anticipation, or "correction" mechanism, usually the feedback paths, is crucial. The system as a whole can go sluggish on the one hand or oscillatory on the other. Symptoms are inventory chaos, unscheduled overtime, share price volatility, exasperated sales forces, frustrated suppliers, and, worst of all, departing long-time customers. Design questions are what is measured? How is it processed? Where is it sent? And, of course, to what purpose?

Properly designed, feedback control systems determine transient and steady-state performance, reduce delays and resonances, alleviate nonlinearities in the production process, help control product quality, minimize inventory, and alleviate nonlinearities in the production process. By way of explanation, in nonlinear systems, two otherwise independent input changes interact with each other to produce effects different from the sum of the effects of each separately. Understandably, the end results can be confusing if not catastrophic. An example is a negotiated reduction in wages followed immediately by an increase in executive wages. The combination results in a strike; either alone would not.

A second key parameter, the resonance time constant, is a measure of the frequency at which the system or several of its elements tries to oscillate or cycle. Resonances are created in almost every feedback system. The more feedback paths, the more resonances. The business cycle, related to inventory cycling, is one such resonance. Resonances, internal and external, can interact to the point of violent, nonlinear oscillation and destruction, particularly if they have the same or related resonant frequencies. Consequently a warning: **Avoid creating the same resonance time constant in more than one location in a [production] system.**

Delay and resonance times, separately and together, are subject to design. In manufacturing systems, the factors that determine these parameters include inventory size, inventory replacement capacity, information acquisition and processing times, fabrication times, supplier capacity, and management response times. All can have a strong individual and collective influence on such overall system time responses as time to market, material and information flow rates, inventory replacement rate, model change rate, and employee turnover rate. Few, if any, of these factors can be chosen or designed independently of the rest, especially in complex feedback systems.

Fortunately, there are powerful tools for feedback system design. They include linear transform theory, transient analysis, discrete event mathematics, fuzzy thinking, and some selected nonlinear and time-variant design methods. The properties of at least simple linear systems designed by these techniques can be simulated and adjusted easily. A certain intuition can be developed based upon long experience with them. For example:

- *Behavior with feedback can be very different from behavior without it.*

Positive example: Provide inventory managers with timely sales information and drastically reduce inventory costs. Negative example: Ignore customer feedback and drown in unused inventory.

- *Feedback works. However, the design of the feedback path is critical. Indeed, in the case of strong feedback, its design can be more important than that of the forward path.*

Positive example: Customer feedback needs to be supplemented by anticipatory projections of economic trends and of competitor's responses to one's own strategies and actions to avoid delays and surprises. Negative examples: If feedback signals are "out of step" or of the wrong polarity, the system will oscillate, if not go completely out of control. Response that is too little, too late is often worse than no response at all.

- *Strong feedback can compensate for many known vagaries, even nonlinearities in the forward path, but it does so "at the margin."*

Example: Production lines can be very precisely controlled around their operating points; that is, once a desired operating point is reached, tight control will maintain it but off that point or on the way to or from it (e.g., start up, synchronization,[10] and shut down) off-optimum behavior is likely. Example: Just in Time response works well for steady flow and constant volume. It can oscillate if flow is intermittent and volume is small.

- *Feedback systems will inherently resist unplanned or unanticipated change, whether internal or external.* Satisfactory reponses to anticipated changes, however, can usually be assured. *In any case, the response will last at least one time constant (cycle time) of the system.* These properties provide stability against disruption. On the other hand, abrupt mandates, however necessary, will be resisted and the end results may be considerably different in magnitude and timing from what advocates of the change anticipated.

Example: Social systems, incentive programs, and political systems notoriously "readjust" to their own advantage when change is mandated. The resultant system behavior is usually less than, later than, and modified from, that anticipated.

- *To make a major change in performance of a presently stable system is likely to require a number of changes in the overall system design.*

Examples: The change from mass production to lean production to flexible production[11] and the use of robots and high technology.

Not all systems are linear, however. As a warning against overdependence on linear-derived intuition, typical characteristics of nonlinear systems are

- *In general, no two systems of different nonlinearity behave in exactly the same way.*
- *Introducing changes into a nonlinear system will produce different (and probably unexpected) results if they are introduced separately than if they are introduced together.*
- *Introducing even very small changes in input magnitude can produce very different consequences even though all components and processes are deterministic.*

Example: Chaotic behavior (noiselike but with underlying structure) close to system limits is such a phenomenon. Example: When the phone system is saturated with calls and goes chaotic, the planned strategy is to cut off all calls to a particular sector (e.g., California after an earthquake) or revert back to the simplest mode possible (first come, first serve). Sophisticated routing is simply abandoned — it is part of the problem. Example: When a computer abruptly becomes erratic as it runs out of memory, the simplest and usually successful technique is to turn it off and start it up again (reboot), hoping that not too much material has been lost.

- *Noise and extraneous signals irreversibly intermix with and alter normal, intended ones, generally with deleterious results.*

Example: Modification of feedback and control signals is equivalent to modifying system behavior, that is, changing its transient and steady-state behavior. Nonlinear systems are therefore particularly vulnerable to purposeful opposition (jamming, disinformation, overloading).

- *Creating nonlinear systems is of higher risk than creating well-understood, linear ones.* The risk is less that the nonlinear systems will fail under carefully framed conditions than that they will behave strangely otherwise.

Example: In the ultraquality spacecraft business there is an heuristic: **If you can't analyze it, don't build it** — an admonition against unnecessarily nonlinear feedback systems.

The two most common approaches to nonlinearity are, first, when nonlinearities are both unavoidable and undesirable, minimize their effect on end-system behavior through feedback and tight control of operating parameters over a limited operating region. Second, when nonlinearity can improve performance as in discrete and fuzzy control systems, be sure to model and simulate performance *outside* the normal operating range to avoid "nonintuitive" behavior.

The architectural and analytic difficulties faced by modern manufacturing feedback systems are that they are neither simple nor necessarily linear. They are unavoidably complex, probably contain some nonlinearities (limiters and fixed time delays), are multiply interconnected, and are subject to sometimes drastic external disturbances, not the least of which are sudden design changes and shifts in public perception. Their architectures must therefore be robust and yet flexible. Though inherently complex, they must be simple enough to understand and modify at the system level without too many unexpected consequences. In short, they are likely to be prime candidates for the heuristic and modeling techniques of systems architecting.

Lean production

One of the most influential books on manufacturing in recent years is 1990 best seller, *The Machine that Changed the World, The Story of Lean Production.*[12] Although the focus of this extensive survey and study was on automobile production in Japan, the U.S., and Europe, its conclusions are applicable to manufacturing systems in general, particularly the concepts of quality and feedback. A 1994 follow-up book, *Comeback, The Fall & Rise of the American Automobile Industry,*[13] tells the story of the American response and the lessons learned from it. The story of lean production systems is by no means neat and orderly. Although the principles can be traced back to 1960, effective implementation took decades. Lean production certainly did not emerge full blown. Ideas and developments came from many sources, some prematurely. Credits were sometimes misattributed. Many contributors were very old by the time their ideas were widely understood and applied. Quality was sometimes seen as an end result instead of as a prerequisite for any change. The remarkable fact that virtually *every* critical parameter improved by at least 20%, if not 50%,[14] does not seem to have been anticipated. Then, within a few years, everything worked. But when others attempted to adopt the process, they often failed. Why?

One way to answer such questions is to diagram the lean production process from an architectural perspective. Figure 4.4 is an architect's sketch of the lean production waterfall derived from the texts of the just-mentioned books, highlighting (bold facing) its nonclassical features and strengthening its classical ones.* The most apparent feature is the number and strength of its feedback paths. Two are especially characteristic of lean production, the supplier waterfall loop, and the customer-sales-delivery loop. Next evident is the Quality Policies box, crucial not only to high quality but to the profitable and proper operation of later steps, Just in Time inventory, rework, and implicit warranties. Quality policies are active elements in the sequence of steps, are a step through which all subsequent orders and specifications must pass, and are as affected by its feedback input as any other step. That is, policies must change with time, circumstance, technology, product change, and process imperatives.

Research and development (R&D) is not "in the loop," but instead is treated as one supplier of possibilities, among many, including one's competitors' R&D. As described in the 1990 study, R&D is not a driver, though it wouldn't be surprising if its future role were different. Strong customer feedback, in contrast, is very much within the loop, making the loop responsive to customer needs at several key points. Manufacturing feedback to suppliers is also a short path, in contrast with the stand-off posture of much U.S. procurement.

The success of lean production has induced mass producers to copy many of its features, not always successfully. Several reasons for lack of success are apparent from the figure. If the policy box does not implement ultraquality, little can be expected from changes further downstream regardless of how much they ought to be able to contribute. Just in Time inventory is an example. Low-quality supply mandates a cushion of inventory roughly proportional to the defect rate; shifting that inventory from the manufacturer back to the supplier without a simultaneous quality upgrade simply increases transportation and financing costs. To

---

* Strictly speaking, though the books' authors did not mention it, an architect's perspective should also include the intersecting product waterfalls and software spirals. Interestingly, because it seems to be true for all successful systems, it is possible to find where and by whom systems architecting was performed. Two of the more famous automotive production architects were Taiichi Ohno, the pioneer of the Toyota Motor Company Production System, and Yoshiki Yamasaki, head of automobile production at Mazda.

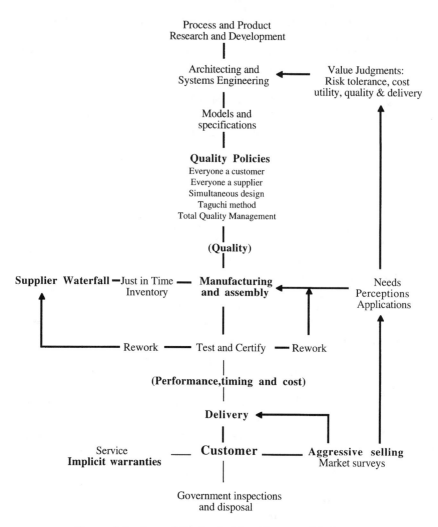

Figure 4.4 An architect's sketch of lean production.

decrease inventory, decrease the defect rate, and *then* apply the coordination principles of JIT, not before.

Another reason for limited success in converting piecemeal to lean production is that any well-operated feedback system, including those used in classical mass production, will resist changes in the forward (waterfall) path. The "loop" will attempt to self-correct. And, it will take at least one loop time constant before all the effects can be seen or even be known. To illustrate, if supply inventory is reduced, what is the effect on sales and service inventory? If customer feedback to the manufacturing line is aggressively sought, as it is in Japan, what is the effect on time-to-market for new product designs?

A serious question raised in both books is how to convert mass production systems into lean production systems. It is not, as the name "lean" might imply, a mass production system with the "fat" of inventory, middle management, screening, and documentation taken out. It is to recognize lean production as a different architecture based on different priorities and interrelationships. How then to begin the conversion? What is both necessary and sufficient? What can be retained?

The one and only place to begin conversion, given the nature of feedback systems, is in the Quality Policies step. In lean production, quality is not a production result determined postproduction and post-test; it is a prerequisite policy imperative. Indeed, Japanese innovators experienced years of frustration when TQM, JIT, and the Taguchi methods at first seemed to do very little. The level of quality essential for these methods to work had not yet been reached. When it was, the whole system virtually snapped into place with results that became famous. Even more important for other companies, unless their quality levels are high enough, even though all the foregoing methods are in place, the famous results will not — and cannot — happen.

Conversely, at an only slightly lower level of quality, lean systems sporadically face at least temporary collapse.[15] As a speculation, there appears to be a direct correlation between how close to the cliff of collapse the system operates and the competitive advantage it enjoys. Backing off from the cliff would seem to decrease its competitive edge, yet getting too close risks imminent collapse — line shutdown, transportation jam-up, short-fuse customer anger, and collateral damage to suppliers and customers for whom the product is an element of a still larger system production.

To summarize, lean production is an ultraquality, dynamic feedback system inherently susceptible to any reduction in quality. It depends upon well-designed, multiple feedback. Given ultraquality standards, lean production arguably is less complex, simpler, and more efficient than mass production. And, by its very nature, it is extraordinarily, fiercely competitive.

Flexible manufacturing

Flexible manufacturing is the capability of sequentially making more than one product on the same production line. In its most common present-day form, it customizes products for individual customers, more or less on demand, by assembling different modules (options) on a common base (platform). Individually tailored automobiles, for example, have been coming down production lines for years. But with one out of three or even one out of ten units having to be sent back or taken out of a production stream, flexible manufacturing in the past has been costly in inventory, complex in operation, and high priced per option compared to all-of-a-kind production.

What has changed is customer demands and expectations, especially in consumer products. Largely because of technological innovation, more capability for less cost now controls purchase rate rather
than wearout and increasing defect rate — an interesting epilog for the
DeSoto story!* One consequence of the change is more models per year
with fewer units per model, the higher costs per unit being offset by
use of techniques such as preplanned product improvement, standardization of interfaces and protocols, and lean production methods.

A natural architecture for the flexible manufacturing of complex
products would be an extension of lean production with its imperatives
— additional feedback paths and ultraquality-produced simplicity —
and an imperative all its own, human-like information comand and
control.

At its core, flexible manufacturing involves the real-time interaction
of a production waterfall with multiple product waterfalls. Lacking an
architectural change from lean production, however, the resultant multiple information flows could overwhelm conventional control systems.
The problem is less that of gathering data than of condensing it. That
suggests that flexible manufacturing will need judgmental, multiple-
path control analogous to that of an air traffic controller in the new
"free flight" regime. Whether the resultant architecture will be fuzzy,
associative, neural, heuristic, or largely human is arguable.

To simplify the flexible manufacturing problem to something more
manageable, most companies today would limit the flexibility to a
product *line* that is forward and backward compatible, uses similar if
not identical modules, keeps to the same manufacturing standards, and
is planned to be in business long enough to write off the cost of the
facility out of product line profits. In brief, production would be limited
to products having a single basic architecture; for example, producing
either Macintosh computers, Hitachi TV sets, or Motorola cellular telephones but not all three on demand on the same production line.

Even that configuration is complex architecturally. To illustrate: A
central issue in product line design is where in the family of products,
from low end to high end, to optimize. Too high in the series, and the
low end is needlessly costly. Too low and the high end adds too little
value. A related issue arises in the companion manufacturing system.
Too much capability and its overhead is too high; too little, and it
sacrifices profit to specialty suppliers.

Another extension from lean production to flexible manufacturing
is much closer coordination between the design and development of the
product line and the design and development of its manufacturing system. Failure to achieve this coordination, as illustrated by the problems

---

* Parenthetically, the Japanese countered the automobile obsolescence problem by
quadrennial government inspections so rigorous that it was often less expensive to
turn a car in and purchase a new one than to bring the old one up to government
standards. Womack et al. 1990, p. 62.

of introducing robots into manufacturing, can be warehouses of unopened crates of robots and in-work products that can't be completed as designed. Clearly: **The product and its manufacturing system must match.** More specifically, their time constants, transient responses, and product-to-machine interfaces must match, recognizing that any manufacturing constraint means a product constraint, and vice versa.

As suggested earlier, the key technological innovation is likely to be the architecture of the joint information system.[16] In that connection, one of the greatest determinates of the information system's size, speed, and resolution is the quality of the end product and the yield of its manufacturing process, that is, their defect rates. The higher these defect rates, the greater the size, cost, complexity, and precision of the information system that will be needed to find and eliminate them quickly.

Another strong determinate of information system capacity is piece part count, another factor dependent on the match of product and manufacturing techniques. Mechanical engineers have known this for years: whenever possible replace a complicated assembly of parts with a single, specialized piece. Nowhere is the advantage of piece part reduction as evident as in the continuing substitution of more and more high-capacity microprocessors for their lower capacity predecessors. Remarkably, this substitution, for approximately the same cost, also decreases the defect rate per computational operation.

Especially for product lines, the fewer different parts from model to model, the better, as long as that commonality doesn't decrease system capability unduly. Once again, there is a close correlation between reduced defect rate, reduced information processing, reduced inventory, and reduced complexity — all by design.

Looking further into the future, an extension of a lean production architecture is not the only possibility for flexible manufacturing. It is possible that flexible manufacturing could take a quite different architectural turn based on a different set of priorities. Is ultraquality production really necessary for simple, low-cost, limited-warranty products made largely from commercial, off-the-shelf (COTS) units, e.g., microprocessors and flat screens? Or is the manufacturing equivalent of parallel processors (pipelines) the answer? Should some flexible manufacturing hark back to the principles of special, handcrafted products one-of-a-kind planetary spacecraft? The answers should be known in less than a decade, considering the profit to be made in finding them.

Heuristics for architecting manufacturing systems

- The product and its manufacturing system must match (in many ways).
- Keep it simple (ultraquality helps).

- Partition for near autonomy (a trade-off with feedback).
- In partitioning a manufacturing system, choose the elements so that they minimize the complexity of material and information flow (Savagian, Peter J., 1990, USC).
- Watch out for critical misfits (between intersecting waterfalls).
- In making a change in the manufacturing process, how you make it is often more important than the change itself (management policy).
- When implementing a change, keep some elements constant to provide an anchor point for people to cling to (Schmidt, Jeffrey H., 1993, USC) (a trade-off when a new architecture is needed).
- Install a machine that even an idiot can use and pretty soon everyone working for you is an idiot (Olivieri, J. M., 1991, USC) (an unexpected consequence of mass production Taylorism). See next heuristic.
- Everyone a customer, everyone a supplier.
- To reduce unwanted nonlinear behavior, linearize!
- If you can't analyze it, don't build it.
- Avoid creating the same resonance time constant in more than one location in a (production) system.
- The five why's (a technique for finding basic causes, and one used by every inquisitive child to learn about the world at large).

In conclusion

Modern manufacturing can be portrayed as an ultraquality, dynamic feedback system intersecting with that of the product waterfall. The manufacturing systems architect's added tasks, beyond those of all systems architects, include (1) maintaining connections to the product waterfall and the software spiral necessary for coordinated developments, (2) assuring quality levels high enough to avoid manufacturing system collapse or oscillation, (3) determining and helping control the system parameters for stable and timely performance, and, last but not least, (4) looking farther into the future than do most product-line architects.

Exercises

1. Manufacturing systems are themselves complex systems which need to be architected. If the manufacturing line is software intensive, and repeated software upgrades are planned, how can certification of software changes be managed?
2. The feedback lags or resonances of a manufacturing system of a commercial product interact with the dynamics of market

demand. Give examples of problems arising from this interaction and possible methods for alleviating them.

3. Examine the following hypothesis: increasing quality levels in manufacturing enables architectural changes in the manufacturing system which greatly increase productivity, but make the system increasingly sensitive to external disruption. For a research exercise, use case studies or a simplified, quantitative model.

4. How does manufacturing systems architecture differ in mass production systems (thousands of units) and very low quantity production systems (fewer than ten produced systems)?

5. Current flexible manufacturing systems usually build very small lot sizes from a single product line in response to real-time customer demand, for example, an automobile production line that builds each car to order. Consider two alternative architectures for organizing such a system, one using close centralized control and another with fully distributed control based on disseminating customer/supplier relationships to each work cell. What would be advantages and disadvantages of both approaches? How would the architecture of the supporting information systems (extending to sales and customer support) have to differ in the two cases?

Notes and References

1. Hayes, R. H., Wheelwright, S. C., and Clark, K. B., *Dynamic Manufacturing, Creating the Learning Organization*, Free Press, New York, 1988, chap. 7, "The Architecture of Manufacturing: Material and Information Flows", p. 185, defines a manufacturing architecture as including its hardware, material, and information flows, their coordination, and managerial philosophy. This textbook is highly recommended, especially Chapters 7, 8, and 9.
2. Rechtin, E., *Systems Architecting, Creating & Building Complex Systems*, Prentice-Hall, Englewood Cliffs, NJ, 1991, 160.
3. Hayes, R. H., Wheelwright, S. C., and Clark, K. B., *Dynamic Manufacturing, Creating the Learning Organization*, Free Press, New York, 1988.
4. Womack, J. P., Jones, D. T., and Roos, D., *The Machine That Changed the World, The Story of Lean Production*, Harper Collins Publishers, New York, 1990, Paperback 1991, p. 4.
5. See also Hayes et al., 1988, chapter 8, "Controlling and Improving the Manufacturing Process".
6. Hayes et al., 1988, chap. 7, p. 185.
7. See Brooks, F. P., Jr., No silver bullet, essence and accidents of software engineering, *Computer*, 10, April 1987, especially p. 18; and Simon, H., *The Sciences of the Artificial*, 2nd ed., p. 200, MIT Press, Cambridge, 1987.

8. For the equivalent in a surgical team, see Brooks' *Mythical Man-Month*.
9. See Womak et al., *The Machine that Changed the World, the Story of Lean Production*.
10. See Lindsey, W. C., *Synchronization Systems in Communication and Control*, Prentice-Hall, Englewood Cliffs, NJ, 1972.
11. Hayes et al., 1988, p. 204.
12. Womack et al., 1990.
13. For an update, see Ingrassia, P. and White, J. B., *Comeback, The Fall & Rise of the American Automobile Industry*, Simon & Schuster, New York, 1994.
14. Womack et al., 1994, Title page.
15. Womack et al., 1990, p. 62.
16. Hayes et al., 1988, chap. 7, p. 185.

chapter five

Social systems

Introduction: defining sociotechnical systems

Social: concerning groups of people or the general public

Technical: based on physical sciences and their application

Sociotechnical systems: technical works involving significant social participation, interests, and concerns.

Sociotechnical systems are technical works involving the participation of groups of people in ways that significantly affect the architectures and design of those works. Historically, the most conspicuous have been the large civil works — monuments, cathedrals, urban developments, dams, and roadways among them. Lessons learned from their construction provide the basis for much of civil (and systems) architecting and its theory.[1]

More recently, others, physically quite different, have become much more sociotechnical in nature than might have been contemplated at their inception — ballistic missile defense, air travel, *e-mail*, information networks, welfare, and health delivery, for example. Few can even be conceived, much less built, without major social participation, planned or not. Experiences with them have generated a number of strategies and heuristics of importance to architects in general. Several are presented here. Among them are three heuristics: the four who's, economic value, and the tension between perceptions and facts. In the interests of informed use, as with all heuristics, it is important to understand the context within which they evolved, the sociotechnical domain. Then at the end of the chapter are a number of general heuristics of applicability to sociotechnical systems, in particular.

Public participation

At the highest level of social participation, members of the public directly use — and may own a part of — the system's facilities. At an intermediate level, they are provided a personal service, usually by a public or private utility. Most important, individuals and not the utilities — the architect's clients — are the end users. Examples are highways, communication and information circuits, general aviation traffic control, and public power. Public cooperation and personal responsibility are required for effective operation. That is, users are expected to follow established rules with a minimum of policing or control. Drivers and pilots follow the rules of the road. Communicators respect the twin rights of free access and privacy.

At this level, participating individuals choose and collectively own a major fraction of the system structure — cars, trucks, aircraft, computers, telephones, electrical and plumbing hardware, and so on. In a sense, the public "rents" the rest of the facilities through access charges, fees for use, and taxes. Reaction to a facility failure or a price change tends to be local, quick, and focused. The public's voice is heard through specialized groups such as automobile clubs for highways, retired persons' associations for health delivery, professional societies for power, and communications services and the like. Market forces are used to considerable advantage through adverse publicity in the media, boycotts, and resistance to stock and bond issues, on the one hand, and through enthusiastic acceptance on the other. Recent examples of major architectural changes strongly supported by the public are superhighways, satellite communications, entertainment cable networks, jet aircraft transportation, health maintenance organizations, and a slow shift from polluting fossil fuels to alternative sources of energy.

At the other extreme of social participation are social systems used solely by the government, directly or through sponsorship, for broad social purposes delegated to it by the public; for example, NASA and DoD systems for exploration and defense, social security and medicare management systems for public health and welfare, research capabilities for national well-being, and police systems for public safety. The public pays for these services and systems only indirectly, through general taxation. The architect's client and end user is the government. The public's connection with the design, construction, and operation of these systems is sharply limited. Its value judgments are made almost solely through the *political* process, the subject of Chapter 10. They might be best characterized as "politico-technical".

The foundations of sociotechnical systems architecting

The foundations of sociotechnical systems architecting are much the same as for all systems: a systems approach, purpose orientation, modeling, certification, and insight. Social system quality, however, is less a foundation than a case-by-case trade-off; that is, the quality desired depends on the system to be provided. In nuclear power generation, modern manufacturing, and manned space flight, ultraquality is an imperative. But in public health, pollution control, and safety, the level of acceptable quality is only one of many economic, social, political, and technical factors to be accommodated.

However, if sociotechnical systems architecting loses one foundation, ultraquality, it gains another: a professional response to the public's needs and perceptions. Responding to public perceptions is particularly difficult, even for an experienced architect. The public's interests are unavoidably diverse and often incompatible. The groups with the strongest interests change with time, sometimes reversing themselves based on a single incident. Three Mile Island was such an incident for nuclear power utilities. Pictures of the earth from a lunar-bound Apollo spacecraft suddenly focused public attention and support for global environmental management. An election of a senator from Pennsylvania avalanched into widespread public concern over health insurance and medicare systems.

The separation of client and user

In most sociotechnical systems, the client, the buyer of the architect's services, is not the user. This fact can present a serious ethical, as well as technical, problem to the architect: how to treat conflicts between the preferences, if not the imperatives, of the utility agency and those of the public (as perceived by the architect) when preferences strongly affect system design.

It is not a new dilemma. State governments have partly resolved the problem by licensing architects and setting standards for systems which affect the health and safety of the public. Buildings, bridges, and public power systems come to mind. Information systems are on the time horizon. The issuing or denial of licenses helps make sure that public interest comes first in the architect's mind. The setting of standards provides the architect some counterarguments against overreaching demands by the client. But these policies do not treat such conflicts as that of the degree of traffic control desired by a manager of an Intelligent Transporation System (ITS) as compared with that of the individual user/driver, nor the degree of governmental regulation of the Internet to assure a balance of access, privacy, security, profit making, and system efficiency.

One of the ways of alleviating some of these tensions is through economics.

Socioeconomic insights

Social economists bring two special insights to sociotechnical systems. The first insight, which might be called **the four who's,** asks four questions that need to be answered *as a self-consistent set* if the system is to succeed economically; namely, **who benefits? who pays? who provides? and, as appropriate, who loses?**

Example: The answers to these questions of the Bell Telephone System were (1) the beneficiaries were the callers and those that received the calls; (2) the callers paid based on usage because they initiated the calls and could be billed for them; (3) the provider was a monopoly whose services and charges were regulated by public agencies for public purposes; and (4) the losers were those who wished to use the telephone facilities for services not offered or to sell equipments not authorized for connection to those facilities. The telephone monopoly was incentivized to carry out widely available basic research because it generated more and better service at less cost, a result the regulatory agencies desired. International and national security agreements were facilitated by having a single point of contact, the Bell System, for all such matters. Standards were maintained and the financial strategy was long term, typically 30 years. The system was dismantled when the losers evoked antitrust laws, creating a new set of losers, complex billing, standards problems, and a loss of research. Subsequently, separate satellite and cable services were established, further dismantling what had been a single service system.

Example: The answers to the four who's for the privatized Landsat System, a satellite-based optical-infrared surveillance service, were (1) the beneficiaries were those individuals and organizations who could intermittently use high-altitude photographs of the earth; (2) because the *value* to the user of an individual photograph was unrelated to its cost (just as is the case with weather satellite TV), the users could not be billed effectively; (3) the provider was a private profit-making organization which understandably demanded a cost-plus-fixed-fee contract from the government as a surrogate customer; and (4) when the government balked at this result of privatization, the Landsat system faced collapse. Research had been crippled, a French government-sponsored service (SPOT) had acquired appreciable market share, and legitimate customers faced loss of service. Privatization was reversed and the government again became the provider.

Example: Serious debates over the nature of their public health systems are underway in many countries, triggered in large part by the technological advances of the last few decades. These advances have made it possible for humanity to live longer and in better health, but the investments in those gains are considerable. The answers to the four who's are at the crux of the debate. Who benefits — everyone equally at all levels of health? Who pays — regardless of personal health or based on need and ability to pay? Who provides — and determines cost to the user? Who loses — anyone out of work or above some risk level, and who determines who loses?

Regardless of how the reader feels about any of these systems, there is no argument that the answers are matters of great social interest and concern. At some point, if there are to be public services at all, the questions must be answered and decisions made. But who makes them and on what basis? Who and where is "the architect" in each case? How and where is the architecture created? How is the public interest expressed and furthered?

The second economics insight is comparably powerful: **In any resource-limited situation, the true value of a given service or product is determined by what one is willing to give up to obtain it.** Notice that the the subject here is *value,* not price or cost.

Example: The public telephone network provides a good example of the difference between cost and value. The cost of a telephone call can be accurately calculated as a function of time, distance, location (urban or rural), bandwidth, facility depreciation, and so on. But the value depends on content, urgency, priority, personal circumstance, and message type, among other things. As an exercise, try varying these parameters and then estimating what a caller might be willing to pay (give up in recreation or necessity). What is then a fair allocation of costs among all users. Should a sick, remote, poor caller have to pay the full cost of remote TV health service, for example? Should a business which can pass costs on to its customers receive volume discounts for long distance calling via satellite? Should home TV be pay-per-view for everyone? Who should decide on the answers?

These socioeconomic heuristics, used together, can alleviate the inherent tensions among the stakeholders by providing the basis for compromise and consensus among them. The complainants are likely to be those whose payments are perceived as disproportionate to the benefits they receive. The advocates, to secure approval of the system as a whole, must give up or pay for something of sufficient value to the complainants that they agree to compromise. Both need to be made to walk in the other's shoes for a while. And, therein, can be the basis of an economically viable solution.

The interaction between the public and private sectors

A third factor in sociotechnical systems architecting is the strong interplay between the public and private sectors, particularly in the advanced democracies where the two sectors are comparable in size, capability, and influence — but differ markedly in how the general public expresses its preferences.

By the middle of the 1990s, the historic boundaries between public and private sectors in communications, health delivery, welfare services, electric power distribution, and environmental control were in a state of flux. This chapter is not the place to debate the pros and cons. Suffice it to say, the imperatives, interests, and answers to the economists' questions are sharply different in the two sectors.[2] The architect is well advised to understand the imperatives of both sectors prior to suggesting architectures that must accommodate them. For example, the private sector must make a profit to survive; the public sector does not — and treats profits as necessary evils. The public sector must follow the rules; the private sector sees rules and regulations as constraints and deterrents to efficiency. Generally speaking, the private sector does best in providing well-specified things at specified times. The public sector does best at providing services within the resources provided.

Because of these differences, one of the better tools for relieving the natural tension between the sectors is to change the boundaries between them in such negotiable areas as taxation, regulation, services provided, subsidies, billing, and employment. Because perceived values in each of these areas is different in the two sectors and under different circumstances, possibilities can exist where each sector perceives a net gain. The architect's role is to help discover the possibilities, achieve balance through compromise on preferences, and assure a good fit across boundaries.

Architecting a system, such as a public health system, that involves both the public and private sectors can be extraordinarily difficult, particularly if agreement does not exist on a mutually trusted architect, on the answers to the economist's questions, or on the social value of the system relative to that of other socially desireable possibilities.

Facts vs. perceptions: an added tension

Of all the distinguishing characteristics of social systems, the one which most sharply contrasts them with other systems is the tension between facts and perceptions about system behavior. To illustrate the impact on design: architects are well familiar with the trade-offs between performance, schedule, cost, and risk. These competing factors might be thought of as pulling architecting four different directions as sketched in Figure 5.1.

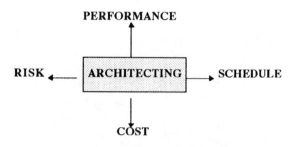

Figure 5.1 Four basic tensions in architecting.

Figure 5.2 can be thought of as the next echelon or ring: the different sources or components of performance, schedule, cost, and risk. Notice that performance has an aesthetic component as well as technical and sociopolitical sources. Automobiles are a clear example. Automobile styling often is more important than aerodynamics or environmental concerns in their architectural design. Costs also have several components, of which the increased costs to people of cost reduction in money and time are especially apparent during times of technological transition. And so on.

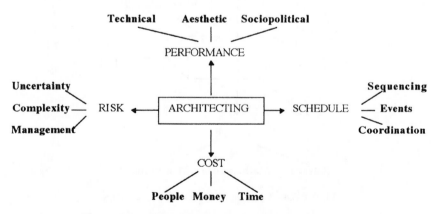

Figure 5.2 Underlying sources of the four tensions.

To these well-known tensions must be added another, one which social systems exemplify but which exist to some degree in all complex systems, namely, the tension between perceptions and facts, shown in Figure 5.3. Its sources are shown in Figure 5.4. This added tension may be dismaying to technically trained architects, but it is all too real to those that deal with public issues. Social systems have generated a painful design heuristic: **It's not the facts, it's the perceptions that count.**

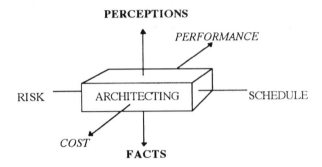

Figure 5.3   Adding facts vs. perceptions.

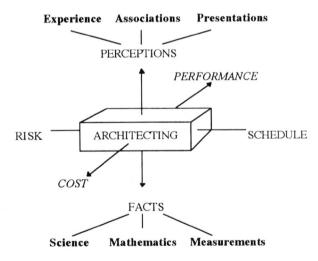

Figure 5.4   Sources of facts and perceptions.

Some real world examples follow:

- It makes little difference what facts nuclear engineers present about the safety of nuclear power plants, their neighbors' perception is that someday the plants will blow up. Remember Three Mile Island and Chernobyl? A.M. Weinberg of Oak Ridge Associated Universities suggested perhaps the only antidote: "The engineering task is to design reactors whose safety is so transparent that the skeptical elite is convinced, and through them the general public."[3]
- Airline travel has been made so safe that the most dangerous part of travel can be driving to and from the airport. Yet, every airliner crash is headline news. A serious design concern, therefore, is how many passengers an airliner should carry — 200? 400? 800? — because even though the average accident rate per departure would probably remain the same,[4] more passsengers would die at once in the larger planes and a public perception might develop that larger airliners are less safe, facts not withstanding.
- One of the reasons that health insurance is so expensive is that health care is perceived by employees as nearly "free" because almost all its costs are paid for either by the employee's company or the government. The facts are that the costs are either passed on to the consumer, subtracted from wages and salaries, taken as a business deduction against taxes, paid for by the general taxpayer, or all of the above. There is no free lunch.
- One of the most profound and unanticipated results of the Apollo flights to the Moon was a picture of the Earth from afar — a beautiful blue, white, brown, and green globe in the blackness of space. We certainly had understood that the Earth was round, but that distant perspective changed our perception of the vulnerability of our home forever, and, with it, our actions to preserve and sustain it. Just how valuable was Apollo, then and in our future? Is there an equivalent value today?

Like it or not, the architect must understand that perceptions can be just as real as facts, just as important in defining the system architecture, and just as critical in determining success. As one heuristic states, **"The phrase, 'I hate it', is direction"**.[5] There have even been times when, in retrospect, perceptions were "truer" than facts which changed with observer, circumstance, technology, and better understanding. Some of the most ironic statements begin with, "It can't be done, because the facts are that..."

Alleviating the tension between facts and perceptions can be highly individualistic. Some individuals can be convinced — in *either* direction — by education, some by prototyping or anecdotes, some by A. M. Weinberg antidote given earlier, some by better packaging or presentation, and some only by the realities of politics. Some individuals will never be convinced, but they might be accepting. In the end, it is a matter of achieving a balance of perceived values. The architect's task is to search out that area of common agreement that can result in a desirable, feasible system.

Heuristics for social systems

- Success is in the eyes of the beholder (not the architect).
- Don't assume that the original statement of the problem is necessarily the best, or even the right one. (Most customers would agree.)
- In conceptualizing a social system, be sure there are mutually consistent answers to the four who's: Who benefits? Who Pays? Who supplies (provides)? And, as appropriate, who loses?
- In any resource-limited situation, the true value of a given service or product is determined by what one is willing to give up to obtain it.
- The choice between the architectures may well depend upon which set of drawbacks the stakeholders can handle best (not on which advantages are the most appealing).
- Particularly for social systems, it's not the facts, it's the perceptions that count. (Try making a survey of public opinion.)
- The phrase "I hate it" is direction (or weren't you listening?).
- In social systems, *how* you do something may be more important than *what* you do (a sometimes bitter lesson for technologists to learn).
- When implementing a change, keep some elements constant as an anchor point for people to cling to (at least until there are some new anchors).
- It's easier to change the technical elements of a social system than the human ones. (Enough said.)

In conclusion

Social systems, in general, place social concerns ahead of technical ones. They exemplify the tension between perception and fact. More than most systems, they require consistent responses to questions of who benefits? who pays? who supplies (provides, builds, etc.), and, sociologically at least, who loses?

Perhaps more than other complex systems, the design and development of social ones should be amenable to insights and heuristics. Social factors, after all, are notoriously difficult to measure, much less predict. But, like heuristics, they come from experience, from failures as well as successes, and from lessons learned.

Exercises

1. Public utilities are examples of sociotechnical systems. How are the heuristics discussed in this chapter reflected in the regulation, design, and operation of a local utility system?

2. Apply the **four who's** to a sociotechnical system familiar to you. Examples: Internet, air travel, communication satellites, a social service.

3. Many efforts are underway to build and deploy intelligent transport systems using modern information technologies to improve existing networks and services. Investigate some of the current proposals and apply the **four who's** to the proposal.

4. Pollution and pollution control are examples of a whole class of sociotechnical systems where disjunctions in the **four who's** are common. Discuss how governmental regulatory efforts, both through mandated standards and pollution license auctions, attempt to reconcile the **four who's**. To what extent have they been successful? How did you judge success?

5. Among the most fundamental problems in architecting a system with many stakeholders are conflicts in purposes and interests. What architectural options might be used to reconcile them?

6. Give an example of the application of the heuristic, **In introducing technological change,** *how* **you do it is often more important than** *what* **you do.**

Notes and References

1. Lang, J., *Creating Architectural Theory, The Role of the Behavioral Sciences in Environmental Design*, Van Nostrand Reinhold, New York, 1987.
2. Rechtin, E., *Systems Architecting, Creating & Building Complex Systems*, Prentice-Hall, Englewood Cliffs, NJ, 1991, p. 270; and "Why not more straight commercial buying" *Government Executive*, October 1976, p. 46.
3. Weinberg, A. M. Engineering in an age of anxiety: the search for inherent safety, in *Engineering and Human Welfare NAE 25, Proc. 25th Annual Meeting*, National Academy of Engineering, Washington, D.C.
4. U.S. airline safety, scheduled commercial carriers, in *The World Almanac® and Book of Facts 1994*, Funk and Wagnalls Corporation © 1993. According to the National Transportation Safety Board source, the fatal accidents per 100,000 mi have remained at or less than 0.100 from 1977 through 1992 despite a 40% increase in departures, major technological change, and the addition of increasingly larger aircraft to the airline fleets.
5. Gradous, L. I., USC, October 18, 1993.

chapter six

Software systems

*Today I am more convinced than ever. Conceptual integrity
is central to product quality. Having a system architect is
the most important step toward conceptual integrity.*

Frederick P. Brooks, Jr., The Mythical Man-Month
After Twenty Years

Introduction: the status of software architecting

Software is rapidly becoming the centerpiece of complex system
design, in the sense that an increasing fraction of system performance
and complexity is captured in software. Software is increasingly the
portion of the system that enables the unique behavioral characteristics
of the system. Competitive developers of end-user system products
find themselves increasingly developing software, even though the
system itself combines both hardware and software. The reasons stem
from software's ability to create intelligent behavior and quickly to
accommodate technical-economic trends in hardware development.

The focus of this chapter is less on the architecting of software
(though that is discussed here and in Part 3) than it is on the impact
of software on system architecting. Software possesses two key
attributes that affect architecting. First, well-architected software can
be very rapidly evolved. Evolution of deployed software is much more
rapid than of deployed hardware because an installed base of software
can be regularly replaced; annual and even quarterly replacement is
common. Annual, or more frequent, field software upgrades are normal
for operating systems, data bases, end-user business systems, large-
scale engineering tools, and communication and manufacturing systems. This puts a demanding premium on software architectures
because they must be explicitly designed to accommodate future
changes and to allow repeated certification with those changes.

Second, software is an exceptionally flexible medium. Software can
easily be built which embodies many logically complex concepts such
as layered languages, rule-driven execution, data relationships, and

many others. This flexibility of expression makes software an ideal medium with which to implement system "intelligence". In both the national security and commercial worlds, intelligent systems are far more valuable to the user and far more profitable for the supplier than their simpler predecessors.

In addition, a combination of technical and economic trends favor building systems from standardized computer hardware and system-unique software, especially when computing must be an important element of the system. Building digital hardware at very high integration levels yields enormous benefits in cost per gate, but requires comparably large capital investments in design and fabrication systems. These costs are fixed, giving a strong competitive advantage to high production volumes. Achieving high production volumes requires that the parts be general purpose. For a system to reap the benefits of very high integration levels, its developers must either use the standard parts (available to all other developers as well) or be able to justify the very large fixed expense of a custom development. If standard hardware parts are selected, software is the remaining means to provide unique system functionality.

A consequence of software's growing complexity and central role is a recognition of the importance of software architecture and its role in system design. An appreciation of sound architectures and skilled architects is taking hold. The soundness of the software architecture will strongly influence the quality of the delivered system and the ability of the developers to further evolve the system. When a system is expected to undergo extensive evolution after deployment, it is usually more important that the system be easily evolvable than that it be exactly correct at first deployment.

Architects, architectures, and architecting are terms increasingly seen in the technical literature, particularly in the software field.[1] Although the term, software architecture, has not been formally defined, a distillation of commonly used ideas is that the architecture is the overall structure of a software system in terms of components and interfaces. This definition would include the major software components, their interfaces with each other and the outside world, and the logic of their execution (single threaded, interrupted, multi-threaded, combination). A software architectural "style" is seen as a generic framework of components or interfaces that defines a class of software structures. The view taken in this book, and in some of the literature,[2] is somewhat more expansive and includes as well other high-level views of the system: behavior, constraints, applications.

High-level advisory bodies to the Department of Defense are calling for architects of ballistic missile defense, C⁴I (command, control, communications, computers, and intelligence), global surveillance, defense communications, internetted weapon systems, and other "systems-of-systems". Formal standards are in process defining the role, milestones, and deliverables of system architecting.* Many of the ideas and terms of those standards come directly from the software domain, e.g., object-oriented, spiral process model, and rapid prototyping. The carryover should not be a surprise; the systems for which architecting is particularly important are behaviorally complex, data intensive, and software rich. Examples of software-centered systems of similar scope are appearing in the civilian world, such as the Information Superhighway, Internet, global cellular telephone, health care, manned space flight, and flexible manufacturing operations.

The consequences to software design of this accelerating trend to smarter systems are now becoming apparent. For the same reason that guidance and control specialists became the core of systems leadership in the past, software specialists will become the core in the future. In the systems world, software will change from having a support role (usually after the hardware design is fixed) to becoming the centerpiece of complex systems design and operation. As more and more of the behavioral complexity of systems are embodied in software, software will become the driver of system configuration. Hardware will be selected for its ability to support software instead of the reverse. This is now common in business information systems and other applications where compatibility with a software legacy is important.

If software is becoming the centerpiece of system development, it is particularly important to reconcile the demands of system and software development. Even if 90% of the system-specific engineering effort is put into software, the system is still the end product. It is the system, not the software inside, the client wishes to acquire. The two worlds share many common roots, but their differing demands have led them in distinctly different directions. Part of the role of systems architecting is to bring them together in an integrated way.

Software engineering is a rich source for integrated models — models which combine, link, and integrate multiple views of a system. Many of the formalisms now used in systems engineering had their roots in software engineering. This chapter discusses the differences between system architecting and software architecting, current directions in software architecting and architecture, and heuristics and guidelines for software. Chapter 9 provides further detail on three integrated software modeling methods, each aimed at the software component of a different type of system.

---

* The IEEE has recently upgraded a planning group to a working group to develop architecting standards. U.S. Military Standards work is also still in progress. See Chapter 11.

Software as a system component

How does the architecture and architecting of software interact with that of the system as a whole? Software has unique properties that influence overall system structure.

1. Software provides a palette of abstractions for creating system behavior. Software is extensible through layered programming to provide abstracted user interfaces and development environments. Through the layering of software it is possible to directly implement concepts such as relational data, natural language interaction, and logic programming that are far removed from their computational implementation.
2. It is economically and technically feasible to use evolutionary delivery for software. If architected to allow it, the software component of a deployed system can be completely replaced on a regular schedule.
3. Software cannot operate independently. Software must always be resident on some hardware system and, hence, must be integrated with some hardware system. The interactions between, and integration with, this underlying hardware system become a key element in software-centered system design.

For the moment there are no emerging technologies that are likely to take software's place in implementing behaviorally complex systems. Perhaps some form of biological or nano-agent technology will eventually acquire similar capabilities. In these technologies behavior is expressed through the emergent properties of chaotically interacting organisms. But the design of such a system can be viewed as a form of logic programming, in which the "program" is the set of component construction and interface rules. Then the system, the behavior that emerges from component interaction, is the expression of an implicit program, a highly abstracted form of software.

System architecting adapts to software issues through its models and processes. To take advantage of the rich functionality there must be models that capture the layered and abstracted nature of complex software. If evolutionary delivery is to be successful, and even just to facilitate successful hardware/software integration, the architecture must reconcile continuously changing software with much less frequently changing hardware.

Software for modern systems

Software plays disparate roles in modern systems. Mass market application software, one-of-a-kind business systems, real-time analysis and control software, and human interactive assistants are all software-centered systems, but each is distinct from the other. The software attributes of rich functionality and amenability to evolution match the characteristics of modern systems. These characteristics include:

1. Storage of, and semiautonomous and intelligent interpretation of, large volumes of information
2. Provision of responsive human interfaces that mask the underlying machines and present their operation in metaphor
3. Semiautonomous adaptation to the behavior of the environment and individual users
4. Real-time control of hardware at rates beyond human capability with complex functionality
5. Constructed from mass-produced computing components and system unique software, with the capability to be customized to individual customers
6. Co-evolution of systems with customers as experience with system technology changes perceptions of what is possible

The marriage of high-level language compilers with general purpose computers allows behaviorally complex, evolutionary systems to be developed at reasonable cost. While the engineering costs of a large software system are considerable, they are much less than the costs of developing a pure hardware system of comparable behavioral complexity. Such a pure hardware system could not be evolved without incurring large manufacturing costs once again. Hardware-centered systems do evolve, but at a slower pace. They tend to be produced in similar groups for several years, then make a major jump to new architectures and capabilities. The time of the jump is associated with the availability of new capabilities and the programmatic capability of replacing an existing infrastructure.

Layering of software as a mechanism for developing greater behavioral complexity is exemplified in the continuous emergence of new software languages closer and closer to application domains. The progression of language is from machine level (machine and assembly languages) to general purpose computing (FORTRAN, Pascal, C, C++, Ada), to domain specific (MatLab, Visual Basic for Applications, dBase, SQL, and others). At each level the models are closer to the application, and the language components provide more specific abstractions. By using higher and higher level languages developers are effectively reusing the coding efforts that went into the language's development. Moreover, the new languages provide new computational abstractions or models not immediately apparent in the architecture of the hardware on which the software executes. Consider a logic programming language like PROLOG. A program in PROLOG is more in the nature of hypothesis and theorem proof than arithmetic and logical calculation.

But it executes on a general purpose computer as invisibly as does a FORTRAN program.

Systems, software, and process models

An architectural challenge is to reconcile the integration needs of software and hardware to produce an integrated system. This is both a problem of representation or modeling and of process. Modeling aspects are taken up subsequently in the chapter and in Part 3. On the process side, hardware is best developed with as few iterations as possible, while software can (and often should) evolve through many iterations. Hardware should follow a well-planned design and production cycle to minimize cost, with large-scale production deferred to as close to final delivery as possible (consistent with adequate time for quality assurance). But software cannot be reliably developed without access to the targeted hardware platform for much of its development cycle.

Software distribution costs are comparatively so low that repeated complete replacement of the installed base is normal practice. When software firms ship their upgrades they ship a complete product. Many firms now "ship" patches and limited updates on the Internet eliminating even the cost of media distribution. The cycle of planned replacement is so ingrained that some products (for example, software development tools) are distributed as a subscription: a quarterly CD-ROM with a new version of the product, application notes, documentation, and prerelease components for early review.

In contrast, the costs of hardware are often dominated by the physical production of the hardware. If the system is mass produced this will clearly be the case. Even when production volumes are very low, as in unique customized systems, the production cost is often comparable to or higher than the development cost. As a result it is uneconomic, and hence impractical, to extensively replace a deployed hardware system with a relatively minor modification. Any minor replacement must compete against a full replacement, a replacement with an entirely new system designed to fulfill new or modified purposes.

One important exception to the rule of low deployment costs for software is where the certification costs of new releases are high. For example, one does not casually replace the flight control software of the Space Shuttle any more than one casually replaces an engine. Extensive test and certification procedures are required before a new software release can be used. Certification costs are analogous to manufacturing costs in that they are a cost required to distribute each release but do not contribute to product development.

Waterfalls for software?

For hardware systems, the process model of choice is a waterfall (in one of its pure or modified incarnations). The waterfall model tries to keep iterations local, that is, between adjacent tasks such as requirements and design. Upon reaching production there is no assumption of iteration, except the large-scale iteration of system assessment and eventual retirement and/or replacement. This model fits well within the traditional architecting paradigm as described in Chapter 1.

Software can, and sometimes does, use a waterfall model of development. The literature on software development has long embraced the sequential paradigm of requirements, design, coding, test, delivery. But dissatisfaction with the waterfall model for software led to the spiral model and variants. Essentially all successful software systems are iteratively delivered. Application software iterations are expected as a matter of course. Weapon system and manufacturing software is also regularly updated with refined functionality, new capabilities, and fixes to problems. One reason for software iterations is to fix problems discovered in the field. A waterfall model tries to eliminate such problems by doing a very high quality job of the requirements. Indeed, the success of a waterfall development is strongly dependent on the quality of the requirements. But in some systems the evolvability of software can be exploited to reach the market faster and avoid costly, and possibly fruitless, requirements searches.

> Example: Data communication systems have an effective requirement of interoperating with whatever happens to be present in the installed base. Deployed systems from a global range of companies may not fully comply with published standards, even if the standards are complete and precise (which they often are not). Hence, determining the "real" requirements to interoperate is quite difficult. The most economical way to do so may be to deploy to the field and compile real experience. But that in turn requires that the systems support the ability to determine the cause of interoperation problems and be economically modifiable once deployed to exploit the knowledge gained.

In contrast, a casual attitude toward evolution in systems with safety or mission critical requirements can be tragic.

> Example: The Therac 25 was a software-controlled radiation treatment machine in which software and system failures resulted in six deaths.[3] It was an evolutionary development from a predecessor machine. The evidence suggests that the safety requirements were well understood, but that the system and software architectures both

failed to maintain the properties. The system architecture was flawed in that all hardware safety interlocks (which had been present in the predecessor model) were removed, leaving software checks as the sole safety safeguard. The software architecture was flawed because it did not guarantee the integrity of treatment commands entered and checked by the system operator.

One of the most extensively described software development problems is customized business systems. These are corporate systems for accounting, management, and enterprise-specific operations. They are of considerable economic importance, built in fairly large numbers (though no two are exactly alike), and developed in an environment relatively free of government restrictions. Popular and widely published development methods have strongly emphasized detailed requirements development followed by semi-automated conversion of the requirements to program code, an application-specific waterfall.

While this waterfall is better than *ad hoc* development, results have been disappointing. In spite of years of experience in developing such business systems, large development projects regularly fail. As Tom DeMarco has noted,[4] "somewhere, today, an accounts payable system development is failing" in spite of the thousands of such systems that have been developed in the past. Part of the reason is the relatively poor state of software engineering compared to other fields. Another reason is failure to make effective use of methods known to be useful. An important reason is the lack of an architectural perspective and the benefits it brings.[5]

The architect's perspective is to explicitly consider implementation, requirements, and long-term client needs in parallel. A requirements-centered approach assumes that a complete capture of documentable requirements can be transformed into a satisfactory design. But existing requirements-modeling methods generally fail to capture performance requirements and ill-structured requirements like modifiability, flexibility, and availability. Even where these nonbehavioral requirements are captured, they cannot be transformed into an implementation in an even semi-automated way. And it is the nature of serious technological change that the impact will be unpredictable. As technology changes and experience is gained, what is demanded from systems will change as well.

The spiral model as originally described did not embrace evolution. Its spirals were strictly risk based and designed to lead to a fixed system delivery. Rapid prototyping envisions evolution, but only on a limited scale. Newer spiral model concepts do embrace evolution.[6] Software processes, as implemented, spiral through the waterfall phases, but do so in a sequential approach to moving release levels. This modified model was introduced in Chapter 4, in the context of integrating software with manufacturing systems, and it will be further explored below.

Spirals for hardware?

To use a spiral model for hardware acquisition is equivalent to repeated prototyping. A one-of-a-kind, hardware-intensive system cannot be prototyped in the usual sense. A complete "prototype" is, in fact, a complete system. If it performs inadequately it is a waste of the complete manufacturing cost of the final system. Each one, from the first article, needs to be produced as though it were the only one. A prototype for such a system must be a limited version or component intended to answer specific developmental questions. The development process needs to place strong emphasis on requirements development and attention to detailed purpose throughout the design cycle. Mass-produced systems have greater latitude in prototyping because of the prototype to production cost ratio, but still less than in software. However, the initial "prototype" units still need to be produced. If they are to be close to the final articles they need to be produced on a similar manufacturing line. But setting up a complete manufacturing line when the system is only in prototype stage is very expensive. Setting up the manufacturing facilities may be more expensive than developing the system. As a hardware-intensive system itself, the manufacturing line cannot be easily modified, and leaving it idle while modifying the product to be produced represents a large cost.

Integration: spirals and circles

What process model matches the nature of evolutionary, mixed technology, behaviorally complex systems? As was suggested in Chapter 4, a spiral and circle framework seems to capture the issues. The system should possess stable configurations (represented as circles) and development should iteratively approach those circles. The stable configurations can be software release levels, architectural frames, or hardware configurations.

This process model matches the accepted software practice of moving through defined release levels, with each release produced in cycles of requirements–design–code–test. Each release level is a stable form that is used while the next release is developed. Three types of evolution can be identified. A software product, like an operating system or shrink-wrapped application, has major increments in behavior indicated by changes in the release number, and more minor increments by changes in the number after the "point". Hence a release 7.2 product would be major version seven, second update. The major releases can be envisioned as circles, with the minor releases cycling into them. On the third level are those changes which result in new systems or re-architected old systems. These are conceptually similar to the major releases, but represent even bigger changes. The process with software annotations is illustrated in Figure 6.1. By using a side view one can envision the major releases as vertical jumps. The evolutionary spiral process moves out to the stable major configurations, then jumps up to the next major change. In practice, evolution on one release level may proceed concurrently with development of a major change.

Hardware–software integration adds to the picture. The hardware configurations must also be stable forms, but should appear at different points than the software intermediates on the development timeline. Some stable hardware should be available during software development to facilitate that development. A typical development cycle for an integrated hardware–software system illustrates parallel progressions in hardware and software with each reaching different intermediate stable forms. The hardware progression might be wire-wrap breadboard, production prototype, production, then (possibly) field upgrade. The software moves through a development spiral aiming at a release 1.0 for the production hardware. The number of software iterations may be many more than for the hardware. In late development stages, new software versions may be built weekly.[7] Before that there will normally be partial releases that run on the intermediate hardware forms (the wire-wrap breadboards and the production prototypes). Hardware/software codesign research is working toward environments in which developing hardware can be represented faithfully enough so that physical prototypes are unnecessary for early integration. Such tools may become available, but iteration through intermediate hardware development levels is still the norm in practice.

A related problem in designing a process for integration is the proper use of the heuristic **do the hard part first**. Since software is evolved or iterated this heuristic implies that the early iterations should address the most difficult challenges. Unfortunately, honoring the heuristic is often difficult. In practice the first iterations are often good-looking interface demonstrations or constructs of limited functionality. If interface construction is difficult, or user acceptance of the interface is risky or difficult, this may be a good choice. But if operation to time constraints under loaded conditions is the key problem some other early development strategy should be pursued. In that case the heuristic suggests gathering realistic experimental data on loading and timing conditions for the key processes of the system. Those data can then be used to set realistic requirements for components of the system in its production configuration.

> Example: Call distribution systems manage large numbers of phone personnel and incoming lines as in technical support or phone sales operation. By tying the system into sales data bases it

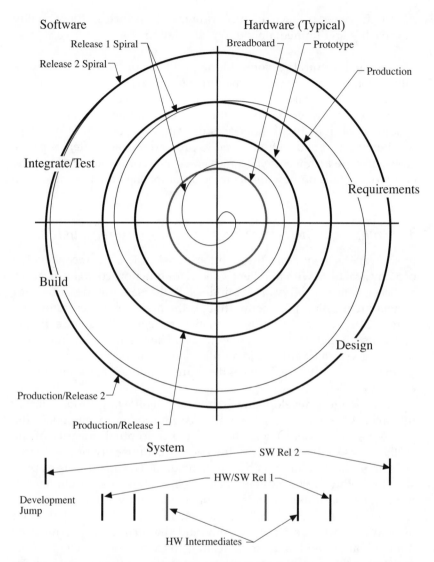

Figure 6.1   A typical arrangement of spirals and circles in a project requiring hardware and software integration. This illustrates the stable intermediate configurations of hardware (typically breadboard and prototype) integrating with a software spiral. Software is developed on the stable intermediate systems.

is possible to develop sophisticated support systems that ensure that full customer information is available in real time to the phone personnel. To be effective the integration of data sources and information handling must be customized to each installation and evolve as understanding of what information is needed and available develops. But, because the system is real time and critical

to customer contact, it must provide its principal functionality
reliably and immediately upon installation.

Thus an architectural response to the problems of hardware–software
integration is to architect both the process and the product. The process
is manipulated to allow different segments of development to match
themselves to the demands of the implementation technology. The
product is designed with interfaces that allow separation of development efforts where the efforts need to proceed on very different paths.
How software architecture becomes an element of system architecture,
and more details on how this is to be accomplished are the subjects to
come.

The role of architecture in software-centered systems

In software as in systems, the architect's basic role is the reconciliation
of a physical form with the client's needs for function, cost, certification,
and technical feasibility. The mind-set is the same as described for
system architecting in general, though the areas of concentration are
different. System architecting heuristics are generally good software
heuristics, though they may be refined and specialized. Several examples are given in Chapter 8. In addition, there are heuristics that apply
particularly to software. Some of these are mentioned at the end of this
chapter.

The architect develops the architecture. Following Brooks' term,[8]
the architect is the user's advocate. As envisioned in this book, the
architect's responsibility goes beyond the conceptual integrity of the
systems as seen by the user, to the conceptual integrity of the system
as seen by the builder and other stakeholders. The architect is responsible for both what the system does as well as how the system does it.
But that responsibility extends, on both counts, only as far as is needed
to develop a satisfactory and feasible system concept. After all, the sum
of both is nearly the whole system, and the architect's role must be
limited if an individual or small team is to carry it out. The latter role,
of defining the overall implementation structure of the system, is closer
to some of the notions of software architecture in recent literature.

The architect's realm is where views and models combine. Where
models that integrate disparate views are lacking, the architect can
supply the insight. When disparate requirements must interact, the
architect's insight can insure that the right characteristics are considered foremost, and, moreover, to develop an architecture that can
reconcile the disparate requirements. The perspective required is predominantly a system perspective. It is the perspective of looking at
the software and its underlying hardware platforms as an integrated
whole which delivers value to the client. Its performance as a whole,
behavioral and otherwise, is what gives it its value.

Architecting for evolution is also an example of the **greatest leverage is at the interfaces** heuristic. Make a system evolvable by paying attention to the interfaces. In software, interfaces are very diverse. With a hardware emphasis, it is common to think of communication interfaces at the bit, byte, or message level. But in software communication interfaces can be much richer and capture extensively structured data, flow of control, and application-specific notions. Much current attention is focused on object interface standards. These are standards for connecting arbitrary software entities together in "documents". Object entities like text, graphics, tables, equations, and spreadsheets seem natural enough. A complete description also requires active objects — often called threads or tasks — which can initiate and control sequences of operation as well as be called by others. The interfaces between software objects can be defined at an essentially arbitrary number of levels of abstraction.

Programming languages, models, and expression

Models are languages. A programming language is a model of a computing machine. Like all languages they have the power to influence, guide, and restrict our thoughts. Programmers with experience in multiple languages understand that some problems will decompose easily in one language, but only with difficulty in another; this is an example of fitting the architecture of the solution to that of a prescriptive solution heuristic. The development of programming languages has been the story of moving successively higher in abstraction from computing hardware.

The layering of languages is essential to complex software development because a high level language is a form of software reuse. Assembly languages masked machine instructions; procedural languages modeled computer instructions in a more language-like prose. Modern languages containing object and strong structuring concepts continue the pattern by providing a richer palette of representation tools for implementing computing constructs. Each statement in FORTRAN, Pascal, or C reuses the compiler writer's machine level implementation of that construct. Even more important examples are the application-specific languages like mathematical languages or data bases. A statement in a mathematical language like MatLab or Mathematica may invoke a complex algorithm requiring long-term development and deep expertise. A data base query language encapsulates complex data storage and indexing code.

One way of understanding this move up the ladder of abstraction is a famous software productivity heuristic on programmer productivity. A purely programming-oriented statement of the heuristic is

> **Programmers deliver the same number of lines of code per day regardless of the language they are writing in.**

Hence, to achieve high software productivity programmers must work in languages that require few lines of code.[9] This heuristic can be used to examine various issues in language and software reuse. The nature of a programming language and the available tools and libraries will determine the amount of code needed for a particular application. Obviously, writing machine code from scratch will require the most code. Moving to high level languages like C or Ada will reduce the amount of original code needed, unless the application is fundamentally one that interacts with the computing hardware at a very low level. Still less original code will be required if the language directly embodies application domain concepts, or, equivalently, application-specific code libraries are available.

Application-specific languages imitate domain language already in use and make it suitable for computing. One of the first and most popular is spreadsheets. The spreadsheet combines a visual abstraction and a computational language suited to a range of modeling tasks in business offices, engineering, and science. An extremely important category is data base query languages. Today it would be quite unusual to undertake an application requiring sophisticated data base functionality and not use an existing data base product and its associated query language. Another more recent category is mathematical languages. These languages, such as Mathematica, MacSyma, and MatLab, use well-understood mathematical syntaxes and then process those languages into computer-processable form. They allow the mathematically literate user to describe solutions in a language much closer to the problem than a general-purpose programming language.

Application-specific programming languages are likely to play an increasingly important role in all systems built in reasonably large numbers. The only impediment to use of these abstractions in all systems is the investment required to develop the language and its associated application generator and tools. One-of-a-kind systems will not usually be able to carry the burden of developing a new language along with a new system unless they fit into a class of system for which a "meta language" exists. Some work along these lines has been done, for example, in command and control systems.[10]

Architectures, "unifying" models, and visions

Architectures in software can be definitions in terms of tasks and modules, language or model constructs, or, at the highest abstraction level, metaphors. Since software is the most flexible and ethereal of media, its architecture, in the sense of a defining structure, can be equally flexible and ethereal.

The most famous example is the Macintosh desktop metaphor, a true architecture. To a considerable degree, when the overall human interface guidelines are added, this metaphor defines the nature of the system. It defines the types of information that will be handled and it defines much of the logic or processing. The guidelines force operation to be human centered; that is, the system continuously parses user actions in terms of the effects on objects in the environment. As a result Macintosh, and now Microsoft Windows, programs are dominated by a main event loop. The foremost structure the programmer must define is the event loop, a loop in which system-defined events are sequentially stripped from a queue, mapped to objects in the environment, and their consequences evaluated and executed.

The power of the metaphor as architecture is twofold. First, the metaphor suggests much that will follow. If the metaphor is a desktop its components should operate similarly to their familiar physical counterparts. This results in fast and retentive learning "by association" to the underlying metaphor. Second, it provides an easily communicable model for the system that all can use to evaluate system integrity. System integrity is being maintained when the implementation to metaphor is clear.

Directions in software architecting

Software architecture and architecting have received considerable recent attention. A special issue of the IEEE Software magazine* has been devoted to "The Artistry of Software Architecture". At least one book covers the subject.[11] Much of the current work in software architecture focuses on architectural structures and their analysis. Much as the term "architectural style" has definite meaning in civil architecture, usage is attached to style in current software work. In the terminology of this book, work on software architecture styles is attempting to find and classify the high level forms of software and their application to particular software problems.

Focusing on architecture is a natural progression of software and programming research that has steadily ascended the ladder of abstraction. Work on structured programming led to structured design and to the multitasking and object-oriented models to be described in Chapter 9. The next stage of the progression is to further classify the large-scale structures that appear as software systems become progressively larger and more complex.

Current work in software architecture primarily addresses the product of architecting (the structure or architecture) rather than the process of generating it. The published studies cover topics such as classifying architectures, mapping architectural styles to particularly appropriate applications, and the use of software frameworks to assemble multiple

---

* *IEEE Software Magazine*, November 1995. This issue contains several articles on software architecture and architecting. It emphasizes the importance of mixing and melding requirements and implementation-driven approaches to producing high quality systems.

related software systems. This book presents some common threads of the architectural process that underlie the generation of architectures in many domains. Once a particular domain such as software is entered, the architect should make full use of the understood styles, frameworks, or patterns in that domain.

The flavor of current work in software architecture is best captured by reviewing some of its key ideas. These include the classification of architectural styles, patterns, and pattern languages in software and software frameworks.

Architectural styles

At the most general level a style is defined by its components, connectors, and constraints. The components are the things from which the software system is composed. The connectors are the interfaces by which the components interact. A style sets the types of components and connectors which will make up the system. The constraints are the requirements which define system behavior. In the current usage, the architecture is the definition in terms of form, which does not explicitly incorporate the constraints. To understand the constraints one must look to additional views.

As a simple example, consider the structured design models described in Chapter 9. A pure structured style would have only one component type, the routine, and only one connector type, invocation with explicit data passing. A software system composed only using these components and connectors could be said to be in the structured style. But the notion of style can be extended to include considerations of its application and deviations.

David Garlan and Mary Shaw give this discussion of what constitutes an architectural style*:

> An architectural style, then defines a family of such systems in terms of a pattern of structural organization. More specifically, an architectural style determines the vocabulary of components and connectors that can be used in instances of that style. Additionally, a style might define topological constraints on architectural descriptions (e.g.. no cycles). Other constraints — say, having to do with execution semantics — might also be part of the style definition.
>
> Given this framework, we can understand what a style is by answering the following questions:

* Garlan, D. and Shaw, M., An Introduction to Software Architecture, Technical Report, School of Computer Science, Carnegie Mellon University, p. 6, 1992.

> What is the structural pattern — the components,
> connectors, and topologies? What is the underly-
> ing computational model? What are the essential
> invariants of the style — its "load bearing walls"?
> What are some common examples of its use?
> What are the advantages and disadvantages of
> using that style? What are some of the common
> specializations?

Garlan and Shaw have gone on to propose several root styles. As an example, their first style is called "pipe and filter". The pipe and filter style contains one type of component, the filter, and one type of connector, the pipe. Each component inputs and outputs streams of data. All filters can potentially operate incrementally and concurrently. The streams flow through the pipes. Likewise, all stream flows are potentially concurrent. Since each component acts to produce one or more streams from one or more streams, it can be thought of as an abstract sort of filter. A pipe and filter system is schematically illustrated in Figure 6.2.

UNIX shell programs and some signal processing systems are common pipe and filter systems. The UNIX shell provides direct pipe and filter abstractions with the filters concurrent UNIX processes and the pipes interprocess communication streams. The pipe and filter abstraction is a natural representation for block-structured signal processing systems in which concurrent entities perform real-time processing on incoming sampled data streams.

Some other styles proposed include object oriented, event based, layered, and blackboard. An object-oriented architecture is built from components that encapsulate both data and function and which exchange messages. An event-based architecture has as its fundamental structure a loop which receives events (from external interfaces or generated internally), interprets the events in the context of system state, and takes actions based on the combination of event and state. Layered architectures emphasize horizontal partitioning of the system with explicit message passing and function calling between layers. Each layer is responsible for providing a well defined-interface to the layer above. A blackboard architecture is built from a set of concurrent components which interact by reading and writing asynchronously to a common area.

Each style carries its advantages and weaknesses. Each of these styles are descriptions of implementations from an implementor's point of view, and specifically from the software implementor's point of view. They are not descriptions from the user's point of view, or even from the point of view of a hardware implementor on the system. A coherent style, at least of the type currently described, gives a conceptual integrity that assists the builder, but may be no help to the user.

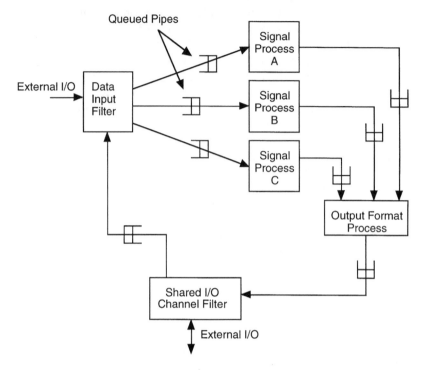

Figure 6.2 A schematic pipe and filter system. Data flow through the system in pipes, which may actually have several types depending on their semantics for queuing, data push or pull, etc. Data are processed in filters which read and write pipes.

Having a coherent implementation style may help in construction, but it is not likely to yield dramatic improvements in productivity or quality because it does not promise to dramatically cut the size of what must be implemented.

The ideal style is one which unifies both the user's and builder's views. The mathematical languages mentioned earlier are examples. They structure the system from both a user's and implementor's point of view. Of course, the internals of the implementation of such a complex software system will contain many layers of abstraction. Almost certainly new styles and abstractions specific to the demands of implementation in real computers will have to arise internally. When ideal styles are not available it is still reasonable to seek models or architectural views which unify some set of considerations larger than just the software implementor. For implementation of complex systems it would be a useful topic of research to find models or styles that encompass a joint hardware–software view.

Architecture through composition

Patterns, styles, and layered abstraction are inherent parts of software practice. Except for the rare machine-level program, all software is built from layered abstractions. High level programming languages impose an intellectual model on the computational machine. The nature of that model inevitably influences what kinds of programs (systems) are built on the machine.

The modern trend is to build systems from components at higher and higher levels of abstraction. It is necessary since no other means is available to build very large and complex systems within acceptable time and effort limits. Each high level library of components imposes its own style and lends itself to certain patterns. The patterns that match the available libraries are encouraged, and it may be very difficult to implement architectures that are not allowed for in the libraries.

> Example: Graphical Macintosh and Windows programs are almost always centrally organized around an event loop and handlers, a type of event-driven style. This structure is efficient because the operating systems provide a built-in event loop to capture user actions such as mouse clicks and key presses. However, since neither has multithreading abstractions (at least before 1995), a concurrent, interacting object architecture is difficult to construct. Many applications would benefit from a concurrent interaction object architecture, but these architectures are very difficult to implement within the constraints of existing libraries.

The logical extension is of higher and higher level languages and libraries to application-specific languages that directly match the nature of the problem they were meant to solve. The modern mathematical software packages are, in effect, very high level software languages designed to mimic the problem they are meant to solve. The object of the packages is to do technical mathematics. So rather than provide a language into which the scientist or engineer must translate mathematics, the package does the mathematics. This is similar for computer-aided design packages, indeed most of the shrink-wrap software industry. These packages surround the computer with a layered abstraction that closely matches the way users are already accustomed to working.

Actually, the relationship between application-specific programming language, software package, and user is more symbiotic. Programmers adapt their programs to the abstractions familiar to the users. But users eventually adapt their abstractions to what is available and relatively easy to implement. The best example is the spreadsheet. The spreadsheet as an abstraction partially existed in paper form as the general ledger. The computer-based abstraction has proven so logical that users have adapted their thinking processes to match the structure of spreadsheets. It should probably be assumed that his type of inter-active relationship will accelerate when the first generation of children to grow up with computers reaches adulthood.

Heuristics and guidelines in software

The software literature is a rich source for heuristics. Most of those heuristics are specific to the software domain, and are often specific to restricted classes of a software-intensive system. The published sets of software heuristics are quite large. The anniversary edition of Brook's *The Mythical Man-Month: Essays in Software Engineering*[12] includes a new chapter, "The Propositions of the Mythical Man-Month: True or False?" which lists the heuristics proposed in the original work. The new chapter reinforces some central heuristics and rejects a few others as incorrect.

The heuristics given in "Man-Month" are broad ranging, covering management, design, organization, testing, and other topics. Several other sources give specific design heuristics. The best sources are detailed design methodologies that combine models and heuristics into a complete approach to developing software in a particular category or style. Chapter 9 discusses three of the best documented, ADARTS*, structured design**, and object oriented***. Even more specific guide-lines are available for the actual writing of code. A recent book by McConnell[13] contains guidelines for all phases and a detailed bibliog-raphy.

From this large source-set, however, there are a few heuristics that particularly stand out as broadly applicable and as basic drivers for software architecting.

- Choose components so that each can be implemented indepen-dently of the internal implementation of all others.
- Programmer productivity in lines of code per day is largely independent of language. For high productivity use languages as close to the application domain as possible.

---

* The published reference on ADARTS, which is quite thorough, is available through the Software Productivity Consortium, ADARTS Guidebook, SPC-94040-CMC, Version 2.00.13, Vol. 1–2, September 1991. ADARTS is an Ada language-specific method, though its ideas generalize well to other languages. In fact, this has been done, although the resulting heuristics and examples are available only to Software Productivity Consortium members.

** Structured design is covered in many books. The original reference is Yourdon and Constantine, *Structured Design: Fundamentals of a Discipline of Computer Program and Systems Design*, Yourdon Press, 1979.

*** Again, there are many books on object-oriented design, and many controversies about its precise definition and the best heuristics or design rules. The book by Rumbaugh, discussed in Chapter 9, is a good introduction.

- The number of defects remaining undiscovered after a test is proportional to the number of defects found in the test. The constant of proportionality depends on the thoroughness of the test, but is rarely less than 0.5.
- Very low rates of delivered defects can be achieved only by very low rates of defect insertion *throughout* software development, and by layered defect discovery — reviews, unit test, system test.
- Software should be grown or evolved, not built.
- The cost of removing a defect from a software system rises exponentially with the number of development phases since the defect was inserted.
- The cost of discovering a defect does not rise. It may be cheaper to discover a requirements defect in customer testing than in any other way, hence the importance of prototyping.
- Personnel skill dominates all other factors in productivity and quality.
- Don't fix bugs later, fix them now.

As has been discussed, the evolvability of software is one of its unique attributes. A related heuristic is **A system will develop and evolve much more rapidly if there are stable intermediate forms than if there are not**. In an environment where wholesale replacement is the norm, what constitutes a stable form? The previous discussion has already talked about releases as stable forms and intermediate hardware configurations. From a different perspective the stable intermediate forms are the unchanging components of the system architecture. These fixed elements provide the framework within which the system can evolve. If they are well chosen, that is, if they are conducive to evolution, they will be stable and facilitate further development. A sure sign the architecture has been badly chosen is the need to change it on every major release. The architectural elements involved could be the use of specific data or control structures, internal programming interfaces, or hardware–software interface definitions. Some examples illustrate the impact of architecture on evolution.

> Example: The Point-to-Point Protocol (PPP) is a publicly defined protocol for computer networking over serial connections (such as modems). Its goal is to facilitate broad multivendor interoperability and to require as little manual configuration as possible. The heart of the protocol is the need to negotiate the operating parameters of a changing array of layered protocols (for example, physical link parameters, authentication, IP control, AppleTalk control, compression, and many others). The list of protocols is continuously growing in response to user needs and vendor business perceptions. PPP implements negotiation through a basic state machine that is reused in all protocols, coupled with a framework for structuring

packets. In a good implementation, a single implementation of the state machine can be "cloned" to handle each protocol requiring only a modest amount of work to add each new protocol. Moreover, the common format of negotiations facilitates troubleshooting during test and operation. During the protocol's development the state machine and packet structure have been mapped to a wide variety of physical links and a continuously growing list of network and communication support protocols.

Example: In the original Apple Macintosh operating system the architects decided to not use the feature of their hardware to separate "supervisor" and "user" programs. They also decided to implement a variety of application programming interfaces through access to global variables. These choices were beneficial to the early versions because they improved performance. But these same choices (because of backward compatibility demands) have greatly complicated efforts to implement advanced operating system features such as protected memory and preemptive multitasking. Another architectural choice was to define the hardware–software interface through the Macintosh Toolbox and the published Apple programming guidelines. The combination has proven to be both flexible and stable. It has allowed a long series of dramatic hardware improvements, and now a transfer to a new hardware architecture, with few gaps in backward compatibility (at least for those developers who obeyed the guidelines).

Example: The Internet Protocol combined with the related Transport Control Protocol (TCP/IP) has become the software backbone of the global Internet. Its partitioning of data handling, routing decisions, and flow control has proven to be robust and amenable to evolutionary development. The combination has been able to operate across extremely heterogeneous networks with equipment built by countless vendors. While there are identifiable architects of the protocol suite, control of protocol development is quite distributed with little central authority. In contrast, the proprietary networking protocols developed and controlled by several major vendors have frequently done relatively poorly at scaling to diverse networks. One limitation in the current IP protocol suite that has become clear is the inadequacy of its 32-bit address space. However, the suite was designed from the beginning with the capability to mix protocol versions on a network. As a result, the deployed protocol version has been upgraded several times (and will be again to IPv6) without interfering with the ongoing operation of the Internet.

Exercises

1. Consult one or more of the references for software heuristics. Extract several heuristics and use them to evaluate a software-intensive system.

2. Requirements defects that are delivered to customers are the most costly because of the likelihood they will require extensive rework. But discovering such defects anytime before customer delivery is likewise very costly because only the customer's reaction may make the nature of the defect apparent. One approach to this problem is prototyping to get early feedback. How can software be designed to allow early prototyping and feedback of the information gained without incurring the large costs associated with extensive rework?

3. Pick three software-intensive systems of widely varying scope, for example, a pen computer-based data entry system for warehouses, an internet work communication server, and the flight control software for a manned space vehicle. What are the key determinants of success and failure for each system? As an architect, how would these determinants change your approach to concept formulation and certification?

4. Examine some notably successful or unsuccessful software-intensive systems. To what extent was success or failure due to architectural (conceptual integrity, feasibility of concept, certification) issues and to what extent was it due to other software process issues.

5. Are there styles analogous to those proposed for software which jointly represent hardware and software?

Notes and References

1. *IEEE Software Magazine,* Special issue on the The Artistry of Software Architecture, November 1995. This issue contains a variety of papers on software architecture, including the role of architecture in reuse, comparisons of styles, decompositions by view, and building block approaches.
2. Kruchten, P. B., A 4+1 view model of software architecture, *IEEE Software Magazine,* p. 42, November 1995.
3. Leveson, N. G. and Turner, C. S., An investigation of the Therac 25 accidents, *Computer,* p. 18, July 1993.
4. DeMarco and Lister, *Peopleware: Productive Projects and Teams,* Dorset House, New York, 1987.
5. Lambert, B., Beyond Information Engineering: The Art of Information Systems Architecture, Technical Report, Broughton Systems, Richmond, VA, 1994.

6. Software Productivity Consortium, Herndon, VA, Evolutionary Spiral Process, Vol. 1–3, 1994.

7. Maguire, S., *Debugging the Development Process*, Microsoft Press, Redmond, WA, 1994.

8. Brooks, F., *The Mythical Man-Month*, Addison-Wesley, Reading, MA, 1975. This classic book has recently been republished with additional commentary by Brooks on how his observations have held up over time. P. 256 in the new edition.

9. Data associated with this heuristic can be found in several books. Two sources are *Programming Productivity* and *Applied Software Measurement*, both by Capers Jones, McGraw Hill, New York.

10. Doman Specific Software Architectures, USC Information Sciences Institute, Marina Del Rey, CA, Technical Report, 1996.

11. Shaw, M. and Garlan, D., *Software Architecture: Perspectives on an Emerging Discipline*, Prentice-Hall, Englewood Cliffs, NJ, 1996.

12. Brooks, F., *The Mythical Man-Month*, Addison-Wesley, Reading, MA, 1975.

13. McConnell, S., *Code Complete: A Practical Handbook of Software Construction*, Microsoft Press, Redmond, WA, 1993.

Exercise to close part two

Explore another domain much as builder-architected, sociotechnical, manufacturing, and software systems are explored in this chapter. What are the domain's special characteristics? What more broadly applicable lessons can be learned from it? What general heuristics apply to it? Some suggested heuristic domains to explore include:

1. Telecommunications in its several forms: point-to-point telephone network systems, broadcast systems (terrestrial and space), and packet-switched data (the Internet)
2. Electric power, which is widely distributed with collaborative control, is subject to complex loading phenomena (with a social component), and is regulated (References: Hill, David J., *Special Issue on Nonlinear Phenomena in Power Systems: Theory and Practical Implications*, Proc. IEEE, 83, 11, November 1995.)
3. Transportation, in both its current form and in the form of proposed intelligent transportation systems
4. Financial systems, including global trading mechanisms and the operation of regulated economics as a system
5. Space systems, with their special characteristics of remote operation, high initial capital investment, vulnerability to interference and attack, and their effects on the design and operation of existing earth-borne system performing similar functions
6. Existing and emerging media systems, including the collection of competing television systems of private broadcast, public broadcast, cable, satellite, and video recording

Part three

Models and modeling

Introduction to part three

What is the product of an architect? While it is tempting to regard the building or system as the architect's product, the relationship is necessarily indirect; those products are created by the builder. The architect acts to translate between the problem domain concepts of the client and the solution domain concepts of the builder. Great architects go beyond the role of intermediary to make a visionary combination of technology and purpose that exceeds the expectation of builder or client. But the system cannot be built as envisioned unless the architect has a mechanism to communicate the vision and track construction against it. The concrete, deliverable products of the architect, therefore, are models of the system.

Individual models alone are point-in-time representations of a system. Architects need to see and treat each as a member of one of several progressions. The architect's first models define the system concept. As the concept is found satisfactory and feasible, the models progress to the detailed, technology-specific models of design engineers. The architect's original models come into play again when the system must be certified.

A civil architecture analogy

Civil architecture provides a familiar example of modeling and progression. An architect is retained to ensure that the building is pleasing to the client in all senses (aesthetically, functionally, and financially). The product of the architect is intangible; it is the conceptual vision which the physical building embodies and satisfies the client. But the intangible product is worthless without a series of increasingly detailed tangible products, all models of some aspect of the building. Table 1 lists some of the models and their purposes.

The progression of models during the design life cycle can be visualized as a steady reduction of abstraction. Early models may be quite abstract. They may convey only the basic floor plan, associated order-of-magnitude budgets, and renderings encompassing only major aesthetic elements. Early models may cover many disparate designs representing optional building structures and styles. As decisions are made, the range of options narrows and the models become more specific. Eventually the models evolve into construction drawings and itemized budgets and pass into the hands of the builders. As the builders work, models are used to control the construction process and to

Table 1   Models and Purposes in Civil Architecture

Model	Purpose
Physical scale model	Convey look and site placement of building to architect, client, and builder
Floor plans	Work with client to ensure building can perform basic functions desired
External renderings	Convey look of building to architect, client, and builder
Budgets, schedules	Ensure building meets client's financial performance objectives; manage builder relationship
Construction blueprints	Communicate design requirements to builder, provide construction acceptance criteria

ensure the integrity of the architectural concept. Even when the building is finished, some of the models will be retained to assist in future project developments and to act as an as-built record for building alterations.

Guide to part 3

While the form of the models differs greatly from civil architecture to aerospace, computer, or software architectures, their purposes and relationships remain the same. Part 3 discusses the concrete elements of architectural practice, the models of systems, and their development. The discussion is from two perspectives broken into three chapters. First, models are treated as the concrete representations of the various views that define a system. This perspective is treated in general in Chapter 7, and through domain-specific examples in Chapter 9. Second, the evolution and development of models are treated as the core of the architecting process. Chapter 8 develops the idea of progressive design as an organizing principle for the architecting process.

Chapter 7, Representation Models and System Architecting, covers the types of models used to represent systems and their roles in architecting. Because architecting is multidimensional and multidisciplinary, an architecture may require many partially independent views. The chapter proposes a set of six views and reviews major categories of models for each view. Since a coherent and integrated product is the ultimate goal, the models chosen must also be designed to integrate with each other. That is, they must define and resolve their interdependencies and form a complete definition of the system to be constructed.

Chapter 8, Design Progression, looks for principles to organize the eclectic architecting process. A particularly useful principle is that of progression — the idea that models, heuristics, evaluation criteria, and many other aspects of the system evolve on parallel tracks from the abstract to the specific and concrete. Progression also helps tie architecting into the more traditional engineering design disciplines. While this book largely treats system architecting as a general process, independent of domain, in practice it is necessarily strongly tied to individual systems and domains. Nevertheless, each domain contains a core of problems not amenable to rational, mechanistic solution which is closely associated with reconciling customer or client need and with technical capability. This core is the province of architecting. Architects are not generalists, they are system specialists, and their models must refine into the technology-specific models of the domains in which their systems are to be realized.

Chapter 9 returns to models, now tying the previous two chapters together by looking at specific modeling methods. The chapter examines a series of integrating methodologies that illustrate the attributes discussed in the previous chapters: multiple views, integration across views, and progression from abstract to concrete implementation. Examples of integrated models and methods are given for computer-based systems, performance-oriented systems, software-intensive systems, manufacturing systems, and sociotechnical systems. The first part of Chapter 9 describes two general-purpose integrated modeling methods, Hatley-Pirbhai and Quantitative Quality Function Deployment. The former specializes in combining behavioral and physical implementation models, the latter in integrating quantitative performance requirements with behavioral and implementation models. Subsequent sections describe integrated models for software, manufacturing systems, and sociotechnical systems.

Chapter seven

Representation models and system architecting

By relieving the mind of all unnecessary work, a good notation sets it free to concentrate on more advanced problems, and in effect increases the mental power of the [human] race.

Alfred North Whitehead

Introduction: roles, views, and models

Models are the primary means of communication with clients, builders, and users; models are the language of the architect. Models enable, guide, and help assess the construction of systems as they are progressively developed and refined. After the system is built, models, from simulators to operating manuals, help describe and diagnose its operation.

To be able to express system imperatives and objectives and manage system design, the architect should be fluent, or at least conversant, with all the languages spoken in the long process of system development. These languages are those of system specifiers, designers, manufacturers, certifiers, distributors, and users.

The most important models are those which define the critical acceptance requirements of the client and those which define the overall structure of the system. The former are a subset of the entirety of the requirements while the latter are a subset of the complete, detailed system design. Because the architect is responsible for total system feasibility, the critical portions may include highly detailed models of components on which success depends and abstract, top-level models of other components.

Models can be classified by their role or by their content. Role is important in relating models to the tasks and responsibilities not only of architects, but of many others in the development process. Of special importance to architects are modeling methods which tie together otherwise separate models into a consistent whole.

Roles of models

Models fill many roles in system architecting, including:

1. Communication with client, users, and builders
2. Maintenance of system integrity through coordination of design activities
3. Assisting design by providing templates, and organizing and recording decisions
4. Exploration and manipulation of solution parameters and characteristics; guiding and recording aggregation and decomposition of system functions, components, and objects
5. Performance prediction; identification of critical system elements
6. Providing acceptance criteria for certification for use

These roles are not independent; each relates to the other. The foremost role is to communicate. The architect discusses the system with the client, the users (if different), the builders, and possibly many other interest groups. Models of the system are the medium of all such communication. After all, the system itself won't come into being for some time to come! The models used for communication become documentation of decisions and designs and thus vehicles for maintaining design integrity. Powerful, well-chosen models will assist in decision making by providing an evocative picture of the system in development. They will also allow relevant parameters and characteristics to be manipulated and the results seen in terms relevant to client, user, or builder.

Communication with the client has two goals. First, the architect must determine the client's objectives and constraints. Second, the architect must insure that the system to be built reflects the value judgments of the client where perfect fulfillment of all objectives is impossible. The first goal requires eliciting information on objectives and constraints and casting it into forms useful for system design. The second requires that the client perceive how the system will operate (objectives and constraints) and have confidence in the progress of design and construction. In both cases models must be clear and understandable to the client, expressible in the client's own terminology. It is desirable that the models also be expressive in the builder's terms, but because client expressiveness must take priority, proper restatement from client to builder language usually falls to the architect.

User communication is similar to client communication. It requires the elicitation of needs and the comparison of possible systems to meet those needs. When the client is the user this process is simplified. When the client and the users are different (as was discussed in Chapter 5 on sociotechnical systems) their needs and constraints may conflict. The architect is in the position to attempt to reconcile these conflicts.

In two-way communication with the builder, the architect seeks to insure that the system will be built as conceived, that system integrity is maintained. In addition, the architect must learn from the builder those technical constraints and opportunities that are crucial in insuring a feasible and satisfactory design. Models that connect the client and the builder are particularly helpful in closing the iterations from builder technical capability to client objectives.

One influence of the choice of a model set is the nature of its associated "language" for describing systems. Given a particular model set and language it will be easy to describe some types of systems, and awkward to describe others, just as natural languages are not equally expressive of all human concepts. The most serious risk in the choice is that of being blind to important alternate perspectives due to long familiarity (and often success) with models, languages, and systems of a particular type.

Classification of models by view

The terms model, view, and perspective are common in engineering, but their precise meaning is variable from one environment to another. At least two national working groups (IEEE and INCOSE) have struggled with getting broad acceptance of precise definition for these and many other terms. Neither group has reached a settled set as of this writing. For the purposes of this book, view and model are given specific meaning.

A view describes a system with respect to some set of attributes or concerns. The set of views chosen to describe a system is variable. A good set of views should be complete (cover all aspects of a system) and mostly orthogonal (capture different pieces of information). Table 7.1 lists the set of views chosen here as most important to architecting. A system can be "projected" into any view, possibly in several ways. The projection into views, and the collection of models by views is shown schematically in Figure 7.1. Each system has some behavior (abstracted from implementation), has a physical form, retains data, etc. A representation of a system with respect to a particular view is called here a model. Not all views are equally important to system developmental success, nor will the set be constant over time. For example, a system might be behaviorally complex but have relatively simple form. Views that are critical to the architect may play only a supporting role in full development.

Table 7.1   Major System or Architectural Views

Perspective or view	Description
Purpose/objective	What the client wants
Form	What the system is
Behavioral or functional	What the system does
Performance objectives or requirements	How effectively the system does it
Data	The information retained in the system and its interrelationships
Managerial	The process by which the system is constructed and managed

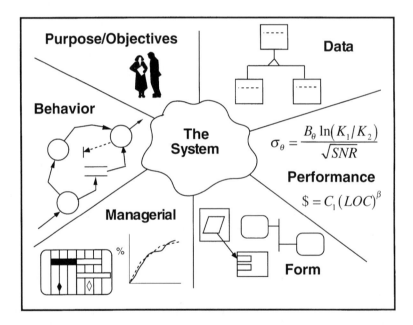

Figure 7.1 Schematic representation of the six views. All views are representations of some aspect of the actual system. Each view may contain several models, as needed to capture the information of the view.

Although any system can be described in each view, the complexity and value of each view's description can differ considerably. Each class of systems emphasizes particular views and has favored modeling methods, or methods of representation within each view. The architect must determine which views are most important to the system and its environment and be expert in the relevant models. While the views are chosen to be reasonably independent, there is extensive linkage among views. For example, the behavioral aspects of the system are not independent of the system's form. The system can produce the desired behavior only if the system's form supports it.

The following sections describe models used for representing a system in each of the views of Table 7.1. The section for each view defines what information is captured by the view, describes the modeling issues within that view, and lists some important modeling methods. Part of the architect's role is to determine which views are most critical to system success, build models for those views, and then integrate as necessary to maintain system integrity. The integration across views is a special concern of the architect.

Note to the reader

The sections to follow, which describe models for each view, are difficult to understand without examples meaningful to each reader. Rather than trying to present detailed examples of each for each of several system domains (a task which might require its own book) we suggest the reader does so on his or her own. The examples given in the chapter are not detailed and are chosen to be understandable to the widest possible audience. Chapter 9 describes, in detail, specific modeling methods that span and integrate multiple views. The methods of Chapter 9 are what the architect should strive for, an integrated picture of all important aspects of a system.

As stated in the introduction, Part 3 can be read several ways. The chapters can be read in order, which captures the intellectual thread of model concepts, modeling processes and heuristics, and specific modeling methods. In this case it is useful to read ahead to Exercises 1 and 2 at the end of this chapter and work them while reading each section to follow. The remaining exercises are intended for after the chapter is read, although some may be approached as each section is completed. An alternative is to read Chapters 7 and 9 in parallel, reading the specific examples of models in Chapter 9 as the views are covered in Chapter 7. Since the approach of Chapter 9 is to look at integrated models, models which span views, a one-for-one correspondence is impossible. The linear approach is probably best for those without extensive background in modeling methods. Those with a good background in integrated modeling methods can use either.

Objectives and purpose models

The first modeling view is that of objectives and purposes. Systems are built for useful purposes, that is, for what the client *wants*. Without them the system cannot survive. The architect's first and most basic role is to match the desirability of the purposes with the practical

feasibility of a system to fulfill those purposes. Given a clearly identifiable client, the architect's first step is to work with that client to
identify the system's objectives and priorities. Some objectives can be
stated and measured very precisely. Others will be quite abstract,
impossible to express quantitatively. Abstract objectives require provisional and exploratory models, models which may fall by the wayside
later as the demands and the resulting system become well understood.
Ideally, all iterations and explorations become part of the system's
document set. However, to avoid drowning in a sea of paper, it may
be necessary to focus on a limited set. If refinement and trade-off
decisions (the creation of concrete objectives from abstract ones) are
architectural drivers, they must be maintained, as it is likely the key
decisions will be repeatedly revisited.

Modeling therefore begins by restating and iterating those initial
unconstrained objectives from the client's language until a modeling
language and methodology emerges, the first major step closer to engineering development. Behavioral objectives are restated in a behavioral
modeling language. Performance requirements are formulated as measurable satisfaction models. Some objectives may translate directly into
physical form, others into patterns of form that should be exhibited by
the system. Complex objectives almost invariably require several steps
of refinement and indeed may evolve into characteristics or behaviors
quite different from their original statement.

A low-technology example (though only by modern standards) is
the European cathedrals of the Middle Ages. A cathedral architect
considered a broad range of objectives. First, a cathedral must fulfill
well-defined community needs. It must accommodate celebration day
crowds, serve as a suitable seat for the Bishop, and operate as a community centerpiece. But, in addition, cathedral clients of that era
emphasized that the cathedral "communicate the glory of God and
reinforce the faithful through its very presence."

Accommodation of holiday celebrants is a matter of size and floor
layout. It is an objective that can be implemented directly and requires
no further significant refinement. The clients — the church and community leaders — because of their personal familiarity with the functioning of a cathedral, could determine for themselves the compliance
of the cathedral by examining the floor plan. But what of defining a
building that "glorifies God?" This is obviously a property only of the
structure as a whole: its scale, mass, space, light, and integration of
decoration and detail. Only a person with an exceptional visual imag-
ination is able to accurately envision the aesthetic and religious impact
of a large structure documented only through drawings and renderings. Especially in those times, when architectural styles were new and
people traveled little, an innovative style would be an unprecedented
experience for perhaps all but the architect.

While refinement of objectives through models is central to architecting, it is also a source of difficulty. A design that proceeds divorced from direct client relevance tends to introduce unnecessary requirements that complicate its implementation. Experience has shown that retaining the client's language throughout the acquisition process can lead to highly efficient, domain-specific architectures, for example, in communication systems.

> Example: Domain Specific Software Architectures[1] are software application generation frameworks in which domain concepts are embedded in the architectural components. The framework is used to generate a product line of related applications in which the client language can be used nearly directly in creating the product. For a set of message handler applications within command and control systems the specification complexity was reduced 50:1.

One measure of the power of a design and implementation method is its ability to retain the original language. But this poses a dilemma. Retention implies the availability of proven, domain-specific methods and engineering tools. But unprecedented systems by definition are likely to be working in new domains, or near the technical frontiers of existing domains. By the very nature of unprecedented system development, such methods and tools are unlikely to be available. Consequently, models and methodologies must be developed and pass through many stages of abstraction, during which the original relevance can easily be lost. The architect must therefore search out domain-specific languages and methods which can somehow maintain the chain of relevance throughout.

An especially powerful, but challenging, form of modeling converts the client/user's objectives into a meta-model or metaphor that can be directly implemented. A famous example is the desktop metaphor adopted for Macintosh computers. The user's objective is to use a computer for daily, office-oriented task automation. The solution is to capture the user's objectives directly by presenting a simulation of a desktop on the computer display. Integrity with user needs is automatically maintained by maintaining the fidelity of a desktop and file system familiar to the user.

Models of form

Models of form represent physically identifiable elements of, and interfaces to, what will be constructed and integrated to meet client objectives. Models of form are closely tied to particular construction technologies, whether the concrete and steel of civil architecture or the less tangible codes and manuals of software systems. Even less tangible physical forms are possible, such as communication protocol standards, a body of laws, or a consistent set of policies.

Models of form vary widely in their degree of abstraction and role. For example, an abstract model may convey no more than the aesthetic feel of the system to the client. A dimensionally accurate but hollow model can assure proper interfacing of mechanical parts. Other models of form may be tightly coupled to performance modeling, as in the scale model of an airplane subjected to wind tunnel testing. The two categories of models of form most useful in architecting are scale models and block diagrams.

Scale models

The most literal of models of form are scale models. Scale models are widely used for client and builder communication, and may function as part of behavioral or performance modeling as well. Some major examples include:

1. Civil architects build literal models of buildings, often producing renderings of considerable artistic quality. These models can be abstracted to convey the feel and style of a building, or can be precisely detailed to assist in construction planning.
2. Automobile makers mock up cars in body-only or full running trim. These models make the auto show circuit to gauge market interest or are used in engineering evaluations.
3. Naval architects model racing yachts to assist in performance evaluation. Scale models are drag tested in water tanks to evaluate drag and handling characteristics. Reduced- or full-scale models of the deck layout are used to run simulated sail handling drills.
4. Spacecraft manufacturers use dimensionally accurate models in fit compatibility tests and in crew extravehicular activity rehearsals. Even more important are ground simulators for on-orbit diagnostics and recovery planning.
5. Software developers use prototypes that demonstrate limited characteristics of a product that are equivalent to scale models; for example, user interface prototypes that looks like the planned system but do not possess full functionality, nonreal-time simulations that carry extensive functionality but do not run in real time, or just a set of screen shots with scenarios for application use.

Physical scale models are gradually being augmented or replaced by virtual reality systems. These "scale" models exist only in a computer and the viewer's mind. They may, however, carry an even stronger impression of reality than a physical scale model because of the sensory immersion achievable.

Block diagrams

A scale model of a circuit board or a silicon chip is unlikely to be of much interest alone, except for expanded scale plots used to check for layout errors. Nonetheless, physical block diagrams are ubiquitous in the electronics industry. To be a model of form, as distinct from a behavioral model, the elements of the block diagram must correspond to physically identifiable elements of the system. Some common types of block diagrams include:

1. System interconnect diagrams that show specific physical elements (modules) connected by physically identifiable channels. On a high-level diagram a module might be an entire computer complex and a channel might be a complex internetwork. On a low level the modules could be silicon chips with specific part numbers and the channels pin-assigned wires.
2. System flow diagrams that show modules in the same fashion as interconnect diagrams but illustrate the flow of information among modules. The abstraction level of information flow defined might be high (complex messaging protocols) or low (bits and bytes). The two types of diagrams (interconnect and flow) are contrasted in Figure 9.3 in Chapter 9.
3. Structure charts,[2] task diagrams,[3] and class and object diagrams[4] that structurally define software systems and map directly to implementation. A software system may have several logically independent diagrams, each showing a different aspect of the physical structure; for example, diagrams that show the invocation tree, the inheritance hierarchy, or the "withing" relationships in a Ada program. Examples of several levels of physical software diagram are given in Figures 9.5 and 9.6 in Chapter 9.
5. Manufacturing process diagrams are drawn with a standardized set of symbols. These represent manufacturing systems at an intermediate level of abstraction, showing specific classes of operation but not defining the machine or the operational details.

Several authors have investigated formalizing block diagrams over a flexible range of architectural levels. The most complete, with widely published examples, is that of Hatley and Pirbhai.[5] Their method is discussed in more depth in Chapter 9 as an example of a method for integrating a multiplicity of architectural views across models. A number of other methods and tools that add formalized physical modeling to behavioral modeling are appearing. Many of these are commercial tools, so the situation is fluid and their methodologies are often not fully defined outside of the tools documentation. Some other examples are the system engineering extensions to ADARTS (described later in the context of software), RDD-100,[6] and StateMate.[7]

Behavioral (functional) models

Functional or behavioral models describe specific patterns of behavior by the system. These are models of what the system *does* (how it behaves) as opposed to what the system *is* (which are models of form). Architects increasingly need behavioral models as systems become more intelligent and their behavior becomes less obvious from the systems form. Unlike a building, a client cannot look at a scale model of a software system and infer how the system behaves. Only by explicitly modeling the behavior can it be understood by the client and builder.

Determining the level of detail or rigor in behavioral specification needed during architecting is an important choice. Too little detail or rigor will mean the client may not understand the behavior being provided (and possibly be unsatisfied) or the builder may misunderstand the behavior actually required. Too much detail or rigor may render the specification incomprehensible, leading to similar problems or unnecessarily delaying development. Eventually, when the system is built, its behavior is precisely specified (if only by the actual behavior of the built system).

From the perspective of architecting, what level of behavioral refinement is needed? The best guidance is to focus on the system acceptance requirements; to ensure the acceptance requirements are passable but complete. Ask what behavioral attributes of the system the client will demand be certified before acceptance and determine through what tests those behavioral attributes can be certified, because the certifiable behavior is the behavior the client will get, no more and no less.

> Example: In software systems with extensive user interface components it has been found by experience that only a prototype of the interface adequately conveys to users how the system will work. Hence, to ensure not just client acceptance, but user satisfaction, an interface prototype should be developed very early in the process. Major office application developers have videotaped office workers as they use prototype applications. The tapes are then examined and scored to determine how effective various layouts were at encouraging users to make use of new features, how rapidly they were able to work, and so on.

> Example: Hardware and software upgrades to military avionics almost always must remain backward compatible with other existing avionics systems and maintain support for existing weapon systems. The architecture of the upgrade must reflect the behavioral requirements of existing system interface. Some may imply very simple behavioral requirements, like providing particular types of

information on a communication bus. Others may demand complex behaviors, such as target handover to a weapon requiring target acquisition, queuing of the weapon sensor, real-time synchronization of the local and weapon sensor feeds, and complex launch codes. The required behavior needs to be captured at the level required for client acceptance, and at the level needed to extract architectural constraints.

Behavioral tools of particular importance are threads or scenarios, data and event flow networks, mathematical systems theory, autonomous system theory, and public choice and behavior models.

Threads and scenarios

A thread or scenario is a sequence of system operations. It is an ordered list of events and actions which represents an important behavior. It normally does not contain branches; that is, it is a single serial scenario of operation, a stimulus/response thread. Branches are represented by additional threads. Behavioral requirements can be of two types. The first type is to require that the system *must* produce a given thread, that is, to require a particular system behavior. The alternative is to require that a particular thread not occur, for example, that a hazardous command never be issued without a positive confirmation having occurred first. The former is more common, but the latter is just as important.

Threads are useful for client communication. Building the threads can be a framework for an interactive dialog with the client. For each input, pose the question "When this input arrives what should happen?" Trace the response until an output is produced. In a similar fashion, trace outputs backward until inputs are reached. The list of threads generated in this way becomes part of the behavioral requirements.

Threads are also useful for builder communication. Even if not complete, they directly convey desired system behavior. They also provide useful tools during design reviews and for test planning. Reviewers can ask that designers walk through their design as it would operate in each of a set of selected threads. This provides a way for reviewers to survey a design using criteria very close to the client's own language. Threads can be used similarly as templates for system tests, ensuring that the tests are directly tied to the client's original dialog.

Data and event flow networks

A complex system can possess an enormous (perhaps infinite) set of threads. A comprehensive list may be impossible, yet without it, the behavioral specification is incomplete. Data and event flow networks allow threads to be collapsed into more compact but complete models. Data flow models define the behavior of a system by a network of functions or processes which exchange data objects. The process network is usually defined in a graphical hierarchy, and most modern versions add some component of finite-state machine description. Current data flow notations are descendants either of DeMarco's data flow diagram (DFD) notation[8] or functional flow block diagrams (FFBD).[9] Chapter 9 gives several examples of data flow models and their relationships with other model types. Figures 9.1 and 9.2 show examples of data flow diagrams for an imaging system. Both the DFD and FFBD methods are based on a set of root principles.

1. The system's functions are decomposed hierarchically. Each function is composed of a network of subfunctions until a "simple" description can be written in text.
2. The decomposition hierarchy is defined graphically.
3. Data elements are decomposed hierarchically and are separately defined in an associated "data dictionary".
4. Functions are assumed to be data triggered. A process is assumed to execute any time its input data elements are available. Finer control is defined by a finite state control model (DFD formalism) or in the graphical structure of the decomposition (FFBD formalism).
5. The model structure avoids redundant definition. Coupled with graphical structuring, this makes the model much easier to modify.

Mathematical systems theory

The traditional meaning of system theory is the behavioral theory of multidimensional feedback systems. Linear control theory is an example of system theory on a limited, well-defined scale. Models of macroeconomic systems and operations research are also system theoretic models, but on a much larger scale.

System theoretic formalisms are built from two components:

1. A definition of the system boundary in terms of observable quantities, some of which may be subject to user or designer control
2. A mathematical machinery that describes the time evolution (the behavior) of the boundary quantities given some initial or boundary conditions and control strategies

There are three main mathematical system formalisms. They are distinguished by how they treat time and data values.

1. Continuous systems: These are the systems classically modeled by differential equations, linear and nonlinear. Values are continuous quantities and are computable for all times.

2. Temporally discrete (sampled data) systems: These are systems with continuously valued elements measured at discrete time points. Their behavior is described by difference equations. Sampled data systems are increasingly important since they are the basis of most computer simulations and nearly all real-time digital signal processing.
3. Discrete event systems: A discrete event system is one in which some or all of the quantities take on discrete values at arbitrary points in time. Queuing networks are the classical example. Asynchronous digital logic is a pure example of a discrete event system. The quantities of interest (say, data packets in a communication network) move around the network in discrete units, but they may arrive or leave a node at an arbitrary, continuous time.

Continuous systems have a large and powerful body of theory. Linear systems have comprehensive analytical and numerical solution methods and an extensive theory of estimation and control. Nonlinear systems are still incompletely understood, but many numerical techniques are available, some analytical stability methods are known, and practical control approaches are available. Similar results are available for sampled data systems. Computational frameworks (based on state machines and Petri Nets) exist for discrete event systems but are less complete than those for differential or difference equation systems in their ability to determine stability and synthesize control laws. A variety of simulation tools are available for all three types of systems. Some tools attempt to integrate all three types into a single framework, though this is difficult.

Many modern systems are a mixture of all three types. For example, consider a computer-based temperature controller for a chemical process. The complete system may include continuous plant dynamics, a sampled data system for control under normal conditions, and discrete event controller behavior associated with threshold crossings and mode changes. A comprehensive and practical modern system theory should answer the classic questions about such a mixed system — stability, closed-loop dynamics, and control law synthesis. No such comprehensive theory exists, but constructing one is an objective of current research. Manufacturing systems are a special example of large-scale mixed systems for which qualitative system understanding can yield architectural guidance.

Autonomous agent, chaotic systems

System-level behavior, as defined in Chapter 1, is behavior not contained in any system component, but which emerges only from the interaction of all the components. A class of system of recent interest is that in which a few types of multiple-replicated, individually relatively simple components interact to create essentially new (emergent) behaviors. Ant colonies, for example, exhibit complex and highly organized behaviors that emerge from the interaction of behaviorally simple, nearly identical, sets of components (the ants). The behavioral programming of each individual ant and its chaotic local interactions with other ants and the environment are sufficient for complex high level behaviors to emerge from the colony as a whole. There is considerable interest in using this truly distributed architecture, but traditional top-down, decomposition-oriented models and their bottom-up integration-oriented complements do not describe it. Some attempts have been made to build theories of such systems from chaos methods. Attempts have also been made to find rules or heuristics for the local intelligence and interfaces necessary for high level behaviors to emerge.

> Example: In some prototype flexible manufacturing plants, instead of trying to solve the very complex work scheduling problem, autonomous controllers schedule through distributed interaction. Each work cell independently "bids" for jobs on its input. Each job moving down the line tries to "buy" the production and transfer services it needs to be completed.[10] Instead of central scheduling the equilibrium of the pseudo-economic bid system distributes jobs and fills work cells. Experiments have shown that rules can be designed that result in stable operation, near optimality of assignment, and very strong robustness to job changes and work cell failure. But the lack of central direction makes it difficult to assure particular operational aspects, for example, to assure that "oddball" jobs won't be ignored for the apparent good of the mean.

Public choice and behavior models

Some systems depend on the behavior of human society as part of the system. In such cases the methods of public choice and consumer analysis may need to be invoked to understand the human system. These methods are often ad hoc, but many have been widely used in marketing analysis by consumer product companies.

> Example: One concept in intelligent transportation systems proposals is the use of centralized routing. In a central routing system each driver would inform the center (via some data network) of his beginning location and his planned destination for each trip. The center would use that information to compute a route for each vehicle and communicate the selected route back to the driver. The route might be dynamically updated in response to accidents or other incidents. In principle, the routing center could adjust routes to optimize the performance of the network as a whole. But would drivers accept centrally selected routes, especially if they thought

the route benefited the network but not them? Would they even bother to send in route information?

A variety of methods could be used to address such questions. At the simplest level are consumer surveys and focus groups. A more involved approach is to organize multiperson driving simulations with the performance of the network determined from individual driver decisions. Over the course of many simulations, as drivers evaluate their own strategies, stable configurations may emerge.

Performance models

A performance model describes or predicts how effectively an architecture satisfies some function. Performance models are usually quantitative, and the most interesting performance models are those of system-level functions, that is, properties possessed by the system as a whole but by no subsystem. Performance models describe properties like overall sensitivity, accuracy, latency, adaptation time, weight, cost, reliability, and many others. Performance requirements are often called "nonfunctional" requirements because they do not define a functional thread of operation, at least not explicitly. Cost, for example, is not a system *behavior*, but it is an important property of the system. Detection sensitivity to a particular signal, however, does carry with it implied functionality. Obviously, a signal cannot be detected unless the processing is in place to produce a detection. It will also usually be impossible to formulate a quantitative performance model without constraining the system's behavior and form.

Performance models come from the full range of engineering and management disciplines. But the internal structure of performance models generally falls into one of three categories:

1. **Analytical:** Analytical models are the products of the engineering sciences. A performance model in this category is a set of lower level system parameters and a mathematical rule of combination that predicts the performance parameter of interest from lower level values. The model is normally accompanied by a "performance budget" or a set of nominal values for the lower level parameters to meet a required performance target.
2. **Simulation:** When the lower level parameters can be identified, but an easily computable performance prediction cannot, a simulation can take the place of the mathematical rule of combination. In essence, a simulation of a system is an analytical model of the systems behavior and performance in terms of the simulation parameters. The connection is just more complex and difficult to explicitly identify. A wide variety of continuous, discrete time, and

discrete event simulators are available, many with rich sets of constructs for particular domains.
3. **Judgmental**: Where analysis and simulation are inadequate or infeasible, human judgment may still yield reliable performance indicators. In particular, human judgment, using explicit or implicit design heuristics, can often rate one architecture as better than another even where a detailed analytical justification is impossible.

Formal methods

The software engineering community has taken a specialized approach to performance modeling known as formal methods. Formal methods seek to develop systems that provably produce formally defined functional and nonfunctional properties. In formal development the team defines system behavior as sets of allowed and disallowed sequences of operation, and may add further constraints, such as timing, to those sequences. They then develop the system in a manner which guarantees compliance to the behavioral and performance definition. Roughly speaking, the formal methods approach is

1. Identify the inputs and outputs of the system. Identify a set of mathematical and logical relations that must exist between the input and output sequences when the system is operating as desired.
2. Decompose the system into components, identifying the inputs and outputs of each component. Determine mathematical relations on each component such that their composition is equivalent to the original set of relations one level up.
3. Continue the process iteratively to the level of primitive implementation elements. In software this would be programming language statements. In digital logic this might be low level combinational or sequential logic elements.
4. Compose the implementation backward up the chain of inference from primitive elements in a way that conserves the decomposed correctness relations. The resulting implementation is then equivalent to the original specification.

From the point of view of the architect, the most important applications of formal methods are in the conceptual phases and in the certification of high-assurance and ultraquality systems. Formal methods require explicit determination of allowed and disallowed input/output sequences. Trying to make that determination can be valuable in eliciting client information, even if the resulting information is not captured in precise mathematical terms. Formal methods also hold out the promise of being able to certify system characteristics that can never be tested. No set of tests can certify with certainty that certain event chains cannot occur, but theorems to that effect are provable within a formal model.

Various formal and semiformal versions of the process are in limited use in software and digital system engineering.[11] While a fully formal version of this process is apparently impractical for large systems at the present time (and is definitely controversial), semiformal versions of the process have been successfully applied to commercial products.

A fundamental problem with the formal methods approach is that the system can never be more "correct" than the original specification. Because the specification must be written in highly mathematical terms it is particularly difficult to use in communication with the typical client.

Data models

The next dimension of system complexity is retained data. What data does the system retain and what relationships among the data does it develop and maintain? Many large corporate and governmental information systems have most of their complexity in their data and its internal relationships. The most common data models have their origins in software development, especially large data base developments. Methods for modeling complex data relationships were developed in response to the need to automate data-intensive, paper-based systems. While data-intensive systems are most often thought of as large, automated data base systems, many working examples are actually paper based. Automating legacy paper-based systems requires capturing the complex interrelationships among large amounts of retained data.

Data models are of increasing importance because of the greater intelligence being embedded in virtually all systems, and the continuing automation of the legacy system. In data-intensive systems, generating intelligent behavior is primarily a matter of finding relationships and imposing persistent structure on the records. This implies that the need to find structure and relationships in large collections of data will be a determinant of system architecture.

> Example: Manufacturing software systems are no longer responsible just for control of work cells. They are part of integrated enterprise information networks in which real-time data from the manufacturing floor, sales, customer operations, and other parts of the enterprise are stored and studied. Substantial competitive advantages accrue to those who can make intelligent judgments from these enormous data sets.

> Example: Intelligent transport systems are a complex combination of distributed control systems, sensor networks, and data

fusion. Early deployment stages will emphasize only simple behavioral adaptation, as in local intelligent light and on-ramp controllers. Full deployment will fuse data sources across metropolitan areas to generate intelligent prediction and control strategies. These later stages will be driven by problems of extracting and using complex relationships in very large data bases.

The basis for modern data models are the entity-relationship diagrams developed for relational data bases. These diagrams have been generalized into a family of object-oriented modeling techniques. An object is a set of "attributes" or data elements and a set of "methods" or functions which act upon the attributes (and possibly other data or objects as well). Objects are instances of classes which can be thought of as templates for specific objects. Objects and classes can have relationships of several types. Major relationship types include aggregation (or composition); generalization, specialization, or inheritance; and association (which may be two-way or M-way). Object-oriented modeling methods combine data and behavioral modeling into a single hierarchy organized along and driven by data concerns. Behavioral methods like those described earlier also include data definitions, but the hierarchy is driven by functional decomposition.

An example of a well-developed, object-oriented data modeling technique (OMT) is given in Chapter 9. Figure 9.7 shows a typical example of the type of diagram used in that method, which combines conventional entity relationship diagram and object-oriented abstraction.

Data-oriented decompositions share the general heuristics of system architecture. The behavioral and physical structuring characteristics have direct analogs — composing or aggregation, decomposition, minimal communications, etc. There are also similar problems of scale. Very large data models must be highly structured with limited patterns of relationship (analogous to limited interfaces) to be implementable.

Managerial models

To both the client and architect, a project may be as much a matter of planning milestones, budgets, and schedules as it is a technical exercise. In sociotechnical systems, planning the system deployment may be more difficult than assembling its hardware. The managerial or implementation view describes the process of building the physical system. It also tracks construction events as they occur.

Most of the models of this view are the familiar tools of project management. In addition, management-related metrics that can be calculated from other models are invaluable in efforts to create an integrated set of models. Some examples include:

1. The waterfall and spiral system development meta-models (They are the templates on which project-specific plans are built.)
2. PERT/CPM and related task and scheduling dependency charts
3. Cost and progress accounting methods
4. Predictive cost and schedule metrics calculable from physical and behavioral models
5. Design or specification time quality metrics — defect counts, postsimulation design changes, rate of design changes after each review

Examples of integrated models

As noted earlier, models which integrate multiple views are the special concern of the architect. These integrating models provide the synthesized view central to the architect's concerns. An integrated modeling method is a system of representation that links multiple views. The method consists of a set of models for a subset of views and a set of rules or additional models to link the core views. Most integrated modeling methods apply to a particular domain. Table 7.2 lists some representative methods. These models are described in greater detail, with examples, in Chapter 9. The references are given there as well.

Table 7.2 Integrated Modeling Methods and Their Domains

Method	Domain	Ref.
Hatley/Pirbhai (H/P)	Computer-based reactive or event-driven systems	Hatley, 1988
Quantitative quality function deployment (Q²FD)	Systems with extensive quantitative performance objectives and understood performance models	Maier, 1995
Object modeling technique (OMT)	Large-scale data-intensive software systems, especially those implemented in modern object languages	Rumbaugh, 1991
ADARTS	Large-scale real-time software systems	SPC, 1991
Manufacturing system analysis (MSA)	Intelligent manufacturing systems	Baudin, 1990

These methods use different models and cover different views. Their components and dimensions are summarized in Table 7.3.

Table 7.3   Comparison of Popular Modeling Methods

View	H/P	OMT	ADARTS	Q²FD	MSA
Objectives	Text	Text	Text	Numbers	Text
Behavior	Data/control flow	Class diagrams, data flow, State charts	Data/event flow	Links only	Data flow
Performance	Text (timing only)	Text	Text	Satisfaction models, QFD matrices	Text, links to standard scheduling models
Data	Dictionary	Class/object diagrams	Dictionary	N/A	Entity-relationship diagrams
Form	Formalized block diagrams	Object diagrams	Task-object-structure charts (multi-level)	Links by allocation	ASME process flow diagrams
Managerial	N/A (link via metrics)	N/A	N/A (link via metrics)	N/A	Funds flow model, scheduling behavior

Summary

An architect's work revolves around models. Since the architect does not build the system directly, its integrity during construction must be maintained through models acting as surrogates. Models will represent and control the specification of the system, its design, and its production plan. Even after the system is delivered, modeling will be the mechanism for assessing system behavior and planning its evolution. Because the architect's concerns are broad, architecting models must encompass all views of the system. The architect's knowledge of models, like an individuals knowledge of language, will tend to channel the directions in which the system develops and evolves.

Modeling for architects is driven by three key characteristics:

1. Models are the principal language of the architect. Their foremost role is to facilitate communication with client and builder. By facilitating communication they carry out their other roles of maintaining design integrity and assisting synthesis.
2. Architects require a multiplicity of views and models. The basic ones are objective, form, behavior, performance, data, and management. Architects need to be aware of the range of models

that are used to describe each of these views within their domain
of expertise, and the content of other views that may become
important in the future.
3. Multidisciplinary, integrated modeling methods tie together the
   various views. They allow the design of a system to be refined
   in steps from conceptually abstract to the precisely detailed necessary for construction.

The next chapter reconsiders the use of modeling in architecture
by placing modeling in a larger set of parallel progressions from
abstract to concrete. There the field of view will expand to the whole
architectural design process and its parallel progressions in heuristics,
modeling, evaluation, and management.

Exercises

1. Choose a system familiar to you. Formulate a model of your
   system in each of the views discussed in the chapter. How effectively does each model capture the system in that view. How
   effectively do the models define the system for the needs of
   initial concept definition and communication with clients and
   builders. Are the models integrated? That is, can you trace information across the models and views?

2. Repeat exercise one, but with a system unfamiliar to you, and
   preferably embodying different driving issues. Investigate models used for the views most unfamiliar to you. In retrospect, does
   your system in Exercise 1 contain substantial complexity in the
   views you are unfamiliar with?

3. Investigate one or more popular computer-aided system or software engineering (CASE) tools. To what extent do they support
   each of the views? To what extent do they allow integration
   across views?

4. A major distinction in behavioral modeling methods and tools
   is the extent to which they support or demand executability in
   their models. Executability demands a restricted syntax and up-
   front decision about data and execution semantics. Do these
   restrictions and demands help or hinder initial concept formulation and communication with builders and clients? If the answer is variable with the system, is there a way to combine the
   best aspects of both approaches?

5. Models of form must be technology specific because they represent actual systems. Investigate modeling formalisms for domains not covered in the chapter; for example, telecommunica-
   tion systems, business information networks, space systems,

integrated weapon systems, chemical processing systems, or financial systems.

Notes and References

1. Balzer, B. and Wile, D., Domain Specific Software Architectures, Technical Reports, Information Sciences Institute, University of Southern California.

2. Yourdon and Constantine, *Structured Design: Fundamentals of a Discipline of Computer Program and Systems Design*, Yourdon Press, Englewood Cliffs, NJ, 1979.

3. ADARTS Guidebook, SPC-94040-CMC, Version 2.00.13, Vol. 1–2, September 1991. Available through the Software Productivity Consortium, Herndon, VA.

4. Rumbaugh, J. et al., *Object-Oriented Modeling and Design*, Prentice-Hall, Englewood Cliffs, NJ, 1991.

5. Hatley, D. J. and Pirbhai, I., *Strategies for Real-Time System Specification*, Dorset House, New York, 1988.

6. A comprehensive system modeling tool marketed by Ascent Logic, Inc.

7. A tool with both discrete event behavioral modeling and physical block diagrams marketed by i-Logix.

8. DeMarco, T., *Structured Analysis and System Specification*, Yourdon Press, Englewood Cliffs, NJ, 1979.

9. Functional Flow Diagrams, AFSCP 375-5 MIL-STD-499, USAF, DI-S-3604/S-126-1, Form DD 1664, June 1968. Much more modern implementations exist, for example, the RDD-100 modeling and simulation tool developed and marketed by Ascent Logic, Inc.

10. Morley, R. E., The chicken brain approach to agile manufacturing, in *Proc. Manufacturing, Engineering, Design, Automation Workshop*, Stanford, Palo Alto, 1992, 19.

11. Two references can be noted; for theory: Hoare, C. A. R., *Communicating Sequential Processes*, Prentice-Hall, Englewood Cliffs, NJ, 1985; for application in software: Mills, H. D., Stepwise refinement and verification in box-structured systems, *IEEE Computer*, p. 23, June 1988.

chapter eight

Design progression in system architecting

Introduction: architecting process components

Having outlined the products of architecting (models) in Chapter 7, this chapter turns to its process. The goal is not a formal process definition. Systems are too diverse to allow a fixed or dogmatic approach to architecting. Instead of trying for a formal process definition, this chapter develops a set of meta-process concepts for architecting activities and their relationships. Architectural design processes are inherently eclectic and wide ranging, going abruptly from the intensely creative and individualistic to the more prescribed and routine. While the processes may be eclectic, they can be organized. Of the various organizing concepts, one of the most useful is stepwise progression or "refinement".

First, a brief review of the architecting process itself: The architect develops system models that span the range of system concerns, from objectives to implementation. The architectural approach is from beginning to end concerned with the feasibility as well as the desirability of the system implementation. It strives for fit, balance, and compromise between client preferences and builder capabilities. Compromise can only be assured by an interplay of activities, including both high-level structuring and such detailed design as is critical to overall success.

This chapter presents a three-part approach to the process of system architecting:

1. An introduction to and review of the general concepts of design, including theories of design, the elements of design, and the processes of creating a design. These frame the activities that make up the progressions.
2. A conceptual model that connects the unstructured processes of architecture to the rigorous engineering processes of the specialty domains or disciplines. This model is based on stepwise reduction of abstraction (or progression) in models, evaluation

criteria, heuristics, and purposes from initial architecting to formal systems engineering.

3. A guide to the place of architecting and its methods with respect to the specialized design domains and the evolutionary development of domain specific methods. Architecting is recursive within a system as it is defined in terms of its implementation domains. A split between architecting and engineering is an irreducible characteristic of every domain, though the boundaries of that split cannot be clear until the scientific basis for the methods in a domain is known.

The progressions of architecting are inextricably bound up with the progressions of all system development. Architecting is not only iterative, it can be recursive. As a system progresses architecting may reoccur on subsystems. The goal here is to understand the intellectual nature of its conduct, whether it happens at a very high level or within a subsystem.

Design concepts for systems architecture

While systems design is an inherently complicated and irregular practice, it has well-established and identifiable characteristics and can be organized into a logical process. As was discussed in the Preface, the activities of architecting can be distinguished from other engineering activities, even if not crisply. Architecting is characterized by the following:

1. Architecting is, predominantly, an eclectic mix of rational and heuristic engineering. Other elements, such as fixed pronouncements and group processes, enter in lesser roles.
2. Architecting revolves around models, but is composed of the basic processes of scoping, aggregation, partitioning, integration, and certification. Few complete rational methods exist for these processes, and the principal guidelines are heuristic.
3. Uncertainty is inherent in complex system design. Heuristics are specialized tools to reduce or control but not eliminate uncertainty.
4. Continuous progression on many fronts is an organizing principle of architecting, architecture models, and supporting activities.

Civil engineering and architecture are perhaps the most mature of all engineering disciplines. Mankind has more experience with engineering civil structures than any other field. If any area could have the knowledge necessary to make it a fully rational and scientific endeavor it should be civil engineering. But it is in civil practice that the distinction between architecture and engineering is best established. Both architects and engineers have their roles, often codified in law, and their professional training programs emphasize different skills. Architects deal particularly with those problems that cannot be entirely rationalized by scientific inquiry. The architect's approach does not ignore science; it combines it with art. Civil engineers must likewise deal with unrationalizable problems, but the focus of their concerns is with well understood, rational design and specification problems. By analogy, this suggests that all design domains contain an irreducible kernel of problems that are best addressed through creative and heuristic approaches that combine art and science.

Historical approaches to architecting

As indicated in the introduction to Part 1, civil architects recognize four basic theories of design: the normative or pronouncement; the rational; the argumentative or consensual; and the heuristic. While all have their roots in the civil architecture practice, they are recognizable in modern complex systems as well. They have been discussed before, in particular in Rechtin 1991,[1] and in the Introduction to Part 1. The purpose in returning to them here is to indicate their relationship to progressive modeling and to bring in their relevance to software-oriented development.*

To review, normative theory is built from pronouncements (statements of what should be), most often given as restrictions on the content of particular views (usually form). A pronouncement demands that a particular type of form be used, essentially unchanged, throughout. Alternatively, one may pronounce the reverse and demand that certain types of form *not* be used. In either case, success is *defined* by accurate implementation of the normative pronouncements, not by measures of fitness.

Rational system design is tightly integrated with modeling since it seeks to derive optimal solutions, and optimality can only be defined within a structured and mathematical framework. To be effective, rational methods require modeling methods which are broad enough to capture all evaluation aspects of a problem, deep enough to capture the characteristics of possible solutions, and mathematically tractable enough to be solved for problems of useful size. Given these, rational methods "mechanically" synthesize a design from a series of modeled problem statements in progressively more detailed subsystems.

Consensual or participative system design uses models primarily as a means of communicating alternative designs for discussion and

---

* Stepwise refinement is a term borrowed from software that describes program development by sequential construction of programs, each complete into itself but containing increasing fractions of the total desired system functionality.

negotiation among participants. From the standpoint of modeling, consensuality is one of several techniques for obtaining understanding and approval of stakeholders, rather than of itself a structured process of design.

General heuristics are guides to — and sometimes obtained from — models, but they are not models themselves. Heuristics are employed at all levels of design, from the most general to domain specific. Heuristics are needed whenever the complexity of the problem, solutions, and issues overwhelms attempts at rational modeling. This occurs as often in software or digital logic design as in general system design. Within a specific domain the heuristic approach can be increasingly formalized, generating increasingly prescriptive guidance. This formalization is a reflection of the progression of all aspects of design — form, evaluation, and emphasis — from abstract to concrete.

The power of the Chapter 2 heuristics comes by cataloging heuristics which apply in many domains, giving them generality in those domains, likely extensibility in others, and a system-level credibility. Applied consistently through the several levels of a system architecture, they help insure system integrity. For example, "Quality cannot be tested in, it must be designed in" is equally applicable from the top-level architectural sketch to the smallest detail. However, the general heuristic relies on an experienced system-level architect to select the ones appropriate for the system at hand, interpret their application specific meaning, and to promulgate them throughout its implementation. A catalog of general heuristics is of much less use to the novice; indeed, an uninformed selection among them could be dangerous. For example, "If it ain't broke, don't fix it," questionable at best, can mislead one from making the small incremental changes that often characterize successful continous improvement programs and can block one from recognizing the qualitative factors, like ultraquality, that redefine the product line completely.

Specialized and formalized heuristics

While there are many very useful general heuristics, there really is not a general heuristic method as such.* Heuristics most often are formalized as part of more formalized methods within specific domains. A formalized heuristic method gives up generality for more direct guidance in its domain. Popular design methods often contain formalized heuristics as guidelines for design synthesis. A good example is the ADARTS** software engineering methodology. ADARTS provides an extensive set of heuristics to transform a data flow-oriented behavioral

---

* On the other hand, knowledge of a codified set of heuristics can lead to new ways of thinking about problems. This could be described as heuristic thinking or qualitative reasoning.

model into a multitasking, modular software implementation. Some examples of formalized ADARTS prescriptive heuristics include:

Map a process to an active I/O process if that transformation interfaces to an active I/O device[2]

Group processes that read or update the same data store or data from the same I/O device into a single process[3]

As the ADARTS method makes clear, these are recommended guidelines and not the success-*defining* pronouncements of the normative approach. These heuristics do not produce an optimal, certifiable, or even unique result, much less success-by-definition. There is ambiguity in their application. Different heuristics may produce conflicting software structurings. The software engineer must select from the heuristic list and interpret the design consequences of a given heuristic with awareness of the specific demands of the problem and its implementation environment.

Conceptually, general and domain-specific formalized heuristics might be arranged in a hierarchy. In this hierarchy, domain-specific heuristics are specializations of general heuristics and the general are abstractions of the specific. Architecting in general and architecting in specific domains may be linked through the progressive refinement and specialization of heuristics. To date, this hierarchy can be clearly identified only for a limited set of heuristics. In any case, the pattern of refining from abstract to specific is a broadly persistent pattern, and is important for understanding life-cycle design progression.

Scoping, aggregation, partitioning, and certification

A development can be envisioned as the creation and transformation of a series of models. For example, to develop the system's requirements is to develop a model of what the system should do and how effectively it should do it. To develop a system design is to develop a model of what the system is. In a pure waterfall development there is rough alignment between waterfall steps and the views defined in Chapter 7. Requirements development develops models for objectives and performance. Functional analysis develops models of behavior, and so on down the waterfall chain. Architects develop models for all views, though not at equal levels of detail. In uncritical views or uncritical portions of the system the models will be rough. In some areas the models may need to be quite detailed from the beginning.

** This method is described in the ADARTS Guidebook, SPC-94040-CMC, Version 2.00.13, Vol. 1, September 1991, available from the Software Productivity Consortium, Herndon, VA.

Models are best understood by view because the views reflect their content. While architecting is consistent with waterfall or spiral development, it does not traverse the steps in the conventional manner. Architects make use of all views and traverse all development steps, but at varying levels of detail and completeness. Since architects tend to follow complex paths through design activities some alternative characterization of design activities independent of view is useful. The principal activities of the architect are scoping, partitioning, aggregation, and certification. Figure 8.1 lists typical activities in each category and suggests some relationships.

Scoping	
Purpose Expansion/Contraction	
Behavioral Definition/Analysis	
Large Scale Alternative Consideration	
Client Satisfaction-Builder Feasibility	
Aggregation	**Partitioning**
Functional Aggregation (abstract)	Behavioral-Functional Decomposition
Functional Aggregation (to physical units)	Physical Decomposition (to lower level design)
Physical components to subsystems	Performance Model Construction
Interface Definition/Analysis	Interface Definition/Analysis
Assembly on timelines or behavioral chains	Decomposition to cyclic processes
Collection into Decoupled Threads	Decomposition into Threads
Certification	
Operational Walkthroughs	
Test and Evaluation	
Verification	
Formal Methods Verification	
Failure Assessment	

Figure 8.1 Typical activities within scoping, aggregation, partitioning, and certification.

Scoping

Scoping procedures are methods for selecting and rejecting problem statements, of defining constraints, and of deciding on what is "inside" or "outside" the system. Scoping implies the ability to rank alternative statements and priorities on the basis of overall desirability or feasibility. Scoping should not design system internals, though some choices may implicitly do so for lack of design alternatives. Desirably, scoping limits what needs to be considered and why.

Scoping is the heart of front-end architecture. A well-scoped system is one which is both desirable and feasible, the essential definition of success in system architecting. As the project begins, the scope forms, at least implicitly. All participants will form mental models of the system and its characteristics; in doing so the system's scope is being defined. If incompatible models appear, scoping has failed through inconsistency. Heuristics suggest that scoping is among the most important of all system design activities. One of the most popular heuristics in Rechtin's book (1991) has been **All the really important mistakes are made the first day**.

Its popularity certainly suggests that badly chosen system scope is a common source for system disasters.

Of course, it is as impossible to prevent mistakes on the first day as it is on any other day. What the heuristic indicates is that mistakes of initial conception will have the worst long-term impact on the project. Therefore one must be particularly careful that a mistake of scope is discovered and corrected as soon as possible.* One way of doing this is to defer absolute decisions on scope by retaining expansive and restrictive options as long as possible, a course of action recommended by another heuristic (the options heuristic of Appendix A).

In principle, scope can be determined rationally through decision theory. Decision theory applies to any selection problem. In this case the things being selected are problem statements, constraints, and system contexts. In practice, the limits of decision theory apply especially strongly to scoping decisions. These limits, discussed in greater detail in a subsequent section, include the problems of utility for multiple system stakeholders, problem scale, and uncertainty. The judgment of experienced architects, at least as expressed through heuristics (see Appendix A for a detailed list), is that the most useful techniques to establish system scope are qualitative.

Scoping heuristics and decision theory share an emphasis on careful consideration of who will use the system and who will judge success. Decision theory requires a utility function, a mathematical representation of system value as a function of its attributes. A utility function can be determined only by knowing whose judgments of system value will have priority and what the evaluation criteria are. Compare the precision of the utility method to related heuristics:

Success is defined by the beholder, not by the architect.

The most important single element of success is to listen closely to what the customer perceives as his requirements and to have the will and ability to be responsive. (J. E. Steiner, 1978)

Ask early about how you will evaluate the success of your efforts. (Hayes-Roth et al., 1983)

Scoping heuristics suggest approaches to setting scope that are outside the usual compromise procedures of focused engineering. One way to resolve intractable problems of scope is to expand. The heuristic is

---

* A formalized heuristic with a similar idea comes from software engineering. It says: **The cost of removing a defect rises exponentially with the time (inproject phases) between its insertion and discovery.** Hence, mistakes of scope (the very earliest) are potentially dominant in defect costs.

Moving to a larger purpose widens the range of solutions.
   (Gerald Nadler, 1990)

The goal of scoping is to form a concept of what the system will
do, how effectively it will do it, and how it will interact with the outside
world. The level of detail required is the level required to gain customer
acceptance, first of continued development and ultimately of the built
system. Thus the scope of the architect's activities are governed not by
the ultimate needs of system development, but by the requirements of
the architect's further role. The natural conclusion to scoping is certification, where the architect determines that the system is fit for use.
Put another way, the certification is that the system is appropriate for
its scope.

Scoping is not solely a requirements-related activity. For scope to
be successfully determined, the resulting system must be both satisfactory and feasible. The feasibility part requires some development of
the system design. The primary activities in design by architects are
aggregation and partitioning, the basic structuring of the system into
components.

Aggregation and partitioning

Aggregation and partitioning are the grouping and separation of
related solutions and problems. They are two sides of the same coin.
Both are the processes by which the system is defined as components.
While one can argue about which precedes the other, in fact, they are
used so iteratively and repeatedly that neither can be usefully said to
precede the other. Conventionally, the components are arranged into
hierarchies with a modest number of components at each level of the
hierarchy (the famous 7 ± 2 structuring heuristic). The most important
aggregation and partitioning heuristics are to minimize external coupling and maximize internal cohesion, usually worded as:[4]

**In partitioning, choose the elements so that they are as indepen-
   dent as possible, that is, elements with low external complexity and high internal cohesion.**

**Group elements that are strongly related to each other, separate
   elements that are unrelated.**

These two heuristics are especially interesting because they are part
of the clearest hierarchy in heuristic refinement. Design is usefully
viewed as progressive or stepwise refinement. Models, evaluation criteria, heuristics, and other factors are all refined as design progresses from
abstract to concrete and specific. Ideally, heuristics exist in hierarchies
that connect general design guidance, such as the two preceding heuristics, to domain-specific design guidelines. The downward direction of

refinement is the deduction of domain-specific guidelines from general heuristics. The upward abstraction is the induction of general heuristics from similar guidelines across domains.

The deductive direction asks, taking the coupling heuristic as an example, how can coupling be measured? Or, for the cohesion heuristic, given alternative designs, which is the most cohesive? Within a specific domain the questions should have more specific answers. For example, within the software domain these questions are answered with greater refinement, though still heuristically. Studies have demonstrated quantitative impact on system properties as the two heuristics are more and less embodied in a systems design. A generally accepted software measure of partitioning is based on interface characterization and has five ranked levels. A related metric for aggregation quality (or cohesion) has seven ranked levels of cohesion.* Studies of large software systems show a strong correlation between coupling and cohesion levels, defect rates, and maintenance costs. A major study[5] found that routines with the worst coupling-to-cohesion ratios (interface complexity to internal coherence) had 7 times more errors and 20 times higher maintenance costs than the routines with the best ratios.

Aggregation and partitioning with controlled fanout and limited communication are a tested approach to building systems in comprehensible hierarchies. Numerous studies in specific domains have shown that choosing loosely coupled and highly cohesive elements leads to systems with low maintenance cost and low defect rates. However, nature suggests that much flatter hierarchies can yield systems of equal or greater robustness, in certain circumstances.

Chapter 7 introduced an ant colony as an example of a flat hierarchy system which exhibits complex and adaptive behavior. The components of an ant colony, a few classes of ants, interact in a very flat system hierarchy. Communication is loose and hierarchically unstructured. There is no intelligent central direction. Nevertheless, the colony as a whole produces complex system-level behavior. The patterns of local, nonhierarchical interaction produce very complex operations. The colony is also very robust in that large numbers of its components (except the queen) can be removed catastrophically and the system will smoothly adapt. A perhaps related technological example is the Internet. Again the system as a whole has a flat hierarchy and no strong central direction. However, the patterns of local communication and resulting collaboration are able to produce complex, stable, and robust system-level behavior.

The observations that controlled and limited fanout and interaction (the 7 ± 2 heuristic and coupling and cohesion studies) and that extreme

---

* The cohesion and coupling levels are carefully discussed in Yourdon, E. and Constantine, L. L., *Structured Design: Fundamentals of a Discipline of Computer Program and Systems Design*, Yourdon Press, Englewood Cliffs, NJ, 1979. They were introduced earlier by the same authors and others in several papers.

fanout and high distributed communication and control (ant colonies and the Internet) can both lead to high quality systems are not contradictory. Rather they are complementary observations of the refinement of aggregation and partitioning into specific domains. In both cases a happy choice of aggregations and partitionings yields good systems. But the specific indicators of what constitutes good aggregation and partitioning vary with the domain. The general heuristic stands for all, but prescriptive or formalized guidance must be adapted for the domain.

Certification

To certify a system is to give an assurance to the paying client that the system is fit for use. Certifications can be elaborate, formal, and very complex, or the opposite. The complexity and thoroughness are dependent on the system. A house can be certified by visual inspection. A computer flight control system might require highly detailed testing, extensive product and process inspections, and even formal mathematical proofs of design elements. Certification presents two distinct problems. The first is determining that the functionality desired by the client and created by the builder is acceptable. The second is the assessment of defects revealed during testing and inspection and the evaluation of those failures with respect to client demands.

Whether or not a system possesses a desired property can be stated as a mathematically precise proposition. Formal methods develop, track, and verify such propositions throughout development, ideally leading to a formal proof that the system as designed possesses the desired properties. However, architecting practice has been to treat such questions heuristically, relying on judgment and experience to formulate tests and acceptance procedures. The heuristics on certification criteria do not address what such criteria should be, but on the process for developing the criteria. Essentially, certification should not be treated separately from scoping or design. Certifiability must be inherent in the design. Two summarizing heuristics — actually on scoping and planning — are

> **For a system to meet its acceptance criteria to the satisfaction of all parties, it must be architected, designed, and built to do so — no more and no less.**

> **Define how an acceptance criterion is to be certified at the same time the criterion is established.**

The first part of certification is intimately connected to the conceptual phase. The system can be certified as possessing desired criteria only to the extent it is designed to support such certification. The second element of certification, dealing with failure, carries its own heuristics. These heuristics emphasize a highly organized and rigorous approach to defect analysis and removal. Once a defect is discovered it should not be considered resolved until it has been traced to its original source, corrected, the correction tested at least as thoroughly as was needed to find the defect originally, and the process recorded. Deming's famous heuristic summarizes:

> **Tally the defects, analyze them, trace them to the source, make corrections, keep a record of what happens afterwards, and keep repeating it.**

A complex problem in ultraquality systems is the need to certify levels of performance that cannot be directly observed. Suppose a missile system is required to have a 99% success rate with 95% confidence. Suppose further that only 50 missiles are fired in acceptance tests (perhaps because of cost constraints). Even if no failures are experienced during testing, the requirement cannot be quantitatively certified. Even worse, suppose a few failures occurred early in the 50 tests but were followed by flawless performance after repair of some design defects. How can the architect certify the system? It is quite possible that the system meets the requirement, but it cannot be proven.

Certification of ultraquality might be deemed a problem of requirements. Many would argue that no requirement should be levied that cannot be quantitatively shown. But the problem will not go away. The only acceptable failure levels in one-of-a-kind systems and those with large public safety impacts will be immeasurable. No such systems can be certified if certification in the absence of quantitatively provable data is not possible.

Some heuristics address this problem. The Deming approach, given as a heuristic above, seeks to achieve any quality level by continuous incremental improvement. Interestingly, there is a somewhat contradictory heuristic in the software domain. When a software system is tested the number of defects discovered should level off as testing continues. The amount of additional test time to find each additional defect should increase and the total number of discovered defects will level out. The leveling out of the number of defects discovered gives an illusion that the system is now defect free. In practice, testing or reviews at any level rarely consistently find more than 60% of the defects present in a system. But if testing at a given level finds only a fixed percentage of defects, it likewise leaves a fixed percentage undiscovered. And the size of that undiscovered set will be roughly proportional to the number found in that same level of test or review. The heuristic can be given in two forms:

> **The number of defects remaining in a system after a given level of test or review (design review, unit test, system test, etc.) is proportional to the number found during that test or review.**

Testing can indicate the absence of defects in a system only when (1) the test intensity is known from other systems to find a high percentage of defects, and (2) few or no defects are discovered in the system under test.

So the discovery and removal of defects are not necessarily an indication of a high quality system. A variation of the "zero-defects" philosophy is that ultraquality requires ultraquality throughout all development processes. That is, certify a lack of defects in the final product by insisting on a lack of defects anywhere in the development process. The ultraquality problem is a particular example of the interplay of uncertainty, heuristics, and rational methods in making architectural choices. That interplay needs to be examined directly to understand how heuristic and rational methods interact in the progression of system design.

Certainty, rationality, and choice

All of the design processes — scoping, partitioning, aggregation, and certification — require decisions. They require decisions on which problem statement to accept, what components to organize the system into, or when the system has reached an acceptable level of development. A by-product of a heuristic-based approach is continuous uncertainty. Looking back on their projects, most architects interviewed for the USC program concluded that the key choices they made were rarely obvious decisions at the time. Although, in retrospect, it may be obvious that a decision was either a very good or a very bad one, at the time the decision was actually made it was not clear at all. The heuristic summarizing is **Before the flight it was opinion, after the flight it was obvious.** The members of the teams were constantly arguing and decision was reached only through the authority of the leading architect for the project.*

A very considerable effort has been made to develop rational decision-making methods. The goal of a fully rational or scientific approach is to make decisions optimally with respect to rigorously determined criteria. Again, the model of architecting practice presented here is a pragmatic mixture of heuristic and rigor. Decision theory works well when the problem can be parameterized with a modest number of values, uncertainty is limited and estimates are reliable, and the client or users possess consistent utility functions with tractable mathematical expression. The absence of any of these conditions weakens or precludes the approach. Unfortunately, some or all of the conditions are usually absent in architecting problems (and even in more restricted disciplinary design problems).

* Comments by Harry Hillaker at USC on his experience as YF-16 architect.

To understand why, one must understand the elements of the decision theoretic approach. The decision theoretic framework is

1. Identify the attributes contributing to client satisfaction and an algorithm for estimating the value of sets of attributes. More formally, this is determining the set over which the client will express preference.
2. Determine a utility function, a function which combines all the attributes and represents overall client satisfaction. Weighted, additive utility functions are commonly used, but not required. The utility function converts preferences into mathematically useful objective function.
3. Include uncertainty by determining probabilities, calculating the utility probability distribution, and determining the client's risk aversion curve. The risk aversion curve is a utility theory quantity that measures the client's willingness to trade risk for return.
4. Select the decision with the highest weighted expected utility.

The first problem in applying this framework to architecting problems is scale. To choose an optimum the decision theory user must be able to maximize the utility functions over the decision set. If the set is very large, the problem is computationally infeasible. If the relationship between the parameters and utility is nonlinear, only relatively small problems are solvable. Unfortunately, both conditions commonly apply to the architecting and engineering of complex systems.

The second problem is to workably and rationally include the effects of uncertainty or risk. In principle, uncertainty and unreliability in estimates can be folded into the decision theoretic framework through probability and assessment of the client's risk aversion curve. The risk aversion curve measures the client's willingness to trade risk and return. A risk-neutral client wants the strategy that maximizes expected return. A risk-averse client prefers a strategy with certainty over opportunity for greater return. A risk-disposed client prefers the opposite, wanting the opportunity for greater return even if the expectation of the return is less.

In practice, however, the process of including uncertainty is heavily subjective. For example, how can one estimate probabilities for unprecedented events? If the probabilities are inaccurate the whole framework loses its claim to optimality. Estimation of risk aversion curves is likewise subjective, at least in practice. When so much subjective judgment has been introduced it is unclear if maintaining the analytical framework leads to much benefit or if it is simply a gloss.

One clear benefit of the decision theory framework is that it makes the decision criteria explicit and, thus, subject to direct criticism and discussion. This beneficial explicitness can be obtained without the full framework. This approach is to drop the analytic gloss, make decisions based on heuristics and architectural judgment, but (and this is more honored in the breach) require the basis be explicitly given and recorded.

A third problem with attempting to fully rationalize architectural decisions is that for many of them there will be multiple clients who have some claim to express a preference. Single clients can be assumed to have consistent preferences and, hence, consistent utility functions. However, consistent utility functions do not generally exist when the client or users is a group as in sociotechnical systems.* Even with single clients, value judgments may change, especially after the system is delivered and the client acquires direct experience.

Rational and analytical methods produce a gloss of certainty, but may hide highly subjective choices. No hard and fast guideline exists for choosing between analytical choice and heuristic choice when unquantified uncertainties exist. Certainly, when the situation is well understood and uncertainties can be statistically measured, the decision theoretic framework is appropriate. When even the right questions are in doubt it adds little to the process to quantify them. Intermediate conditions call for intermediate criteria and methods. For example, a system might have as client objectives "be flexible" and "leave in options." Obviously, these criteria are open to interpretation. The refinement approach is to derive or specialize increasingly specific criteria from very general criteria. This process creates a continuous progression of evaluation criteria from general to specific and eventually measurable.

> Example: The U.S. Department of Transportation has financed an Intelligent Transport System (ITS) architecture development effort. Among their evaluation criteria was "system flexibility", obviously a very loose criterion. An initial refinement of the loose criterion could be
>
> 1. *Architecture components should fit in many alternative architectures.*
> 2. *Architecture components should support multiple services.*
> 3. *Architecture components should expand with linear or sublinear cost to address greater load.*
> 4. *Components should support non-ITS services.*

These refined heuristic evaluation criteria can be applied directly to candidate architectures, or they can be further refined into quantitative and measurable criteria. The intermediate refinement on the way to quantitative and measurable criteria creates a progression that threads through the whole development process. Instead of thinking

---

* This problem with multiple clients and decision theory has been extensively studied in literature on public choice and political philosophy. A tutorial reference is Mueller, D. C., *Public Choice*, Cambridge University Press, New York, 1979.

of design as beginning and stopping, it continuously progresses. Sophisticated mixtures of the heuristics and rationale are part of architecting practice in some domains. This progression is the topic of the next section.

Stopping or progressing?

When does architecting and modeling stop? The short answer is that given earlier, they never stop, they progress. The architecting process (along with many other parallel tracks) continuously progresses from the abstract to the concrete in a steady reduction of abstraction. In a narrow sense, there are recognizable points at which some aspects of architecting and modeling must stop. To physically fabricate a component of a system its design must be frozen. It may not stop until the lathe stops turning or the final line of code is typed in, but the physical object is the realization of *some* design. In the broader sense even physical fabrication does not stop architecting. Operations can be interpreted only through recourse to models, though those models may be quite precise when driven by real data. In some systems, such as distant space probes, even operational modeling is still somewhat remote.

The significant progressions in architecting are promoted by the role of the architect. The architect's role makes two decisions foremost, the selection of a system concept and the certification of the built system. The former decision is almost certain to be driven by heuristic criteria; the latter is more open, depending on how precisely the criteria of fitness for use can be defined. A system concept is suitable when it is both satisfactory and feasible. Satisfactory can be judged only by the client, though the client will have to rely on information provided by the architect. In builder-architected systems the architect must often make the judgment for the client (who will hopefully appear after the system reaches the market). Feasible means the system can be developed and deployed with acceptable risk. Certification requires that the system as built adequately fulfills the client's purposes, including cost, as well as the contract with the builder.

Risk, a principal element in judging feasibility, is almost certain to be judged heuristically. The rational means of handling risk is through probability, but a probabilistic risk assessment requires some set of precedents to estimate over, that is, a series of developments of similar nature for which the performance and cost history is known. By definition such a history cannot be available for unprecedented systems. So the architect is left to estimate risk by other means. In well-defined domains, past history should be able to provide a useful guide; it certainly does for civil architects. Civil architects are expected to control cost and schedule risk for new structures, and can do so because construction cost estimation methods are reasonably well developed. The desired approach is to use judgment, and perhaps a catalog of domain-specific heuristics, to size the development effort against past systems, and to use documented development data from those past systems to estimate risk. For example, in software systems cost models based on code size estimates are known, and are often calibrated against past development projects in builder organizations. If the architect can deduce code size — and possible variation in code size — reliably, a traceable estimate of cost risk is possible.

The judgment of how satisfactory a concept is, and the certification process both depend on how well customer purposes can be specified. Here there is great latitude for both heuristic and rational means. If customer purposes can be precisely specified it may be possible to precisely judge how well a system fulfills them, either in prospect or retrospect. In prospect it depends on having behavior and performance models that are firmly attached to customer purposes. With good models with certain connection between the models and implementation technology, the architect can confidently predict how well the planned system will fulfill the desired purposes. The retrospective problem is that of certification, of determining how well the built system fulfills the customer purposes. Again, well-founded, scientific models and mature implementation technologies make system assessment relatively certain.

More heuristic problems arise when the same factors do not apply. Mainly this occurs when it is hard to formulate precise customer purpose models, when it is hard to determine whether or not a built system fulfills a precisely stated purpose, or when there is uncertainty about the connection between model and implemented system. The second two are related since they both concern retrospective assessment of architecture models against a built system in the presence of uncertainty.

The first case applies when customer purposes are vague or likely to change in response to actual experience with the system. When the customer is relatively inexperienced with systems of the type, his or her perception of the system's value and requirements is likely to change, perhaps radically, with experience. Vague customer purposes can be addressed through the architecture; for example, an emphasis on options in the architecture and a development plan that includes early user prototypes with the ability to feed prototype experience back into the architecture. This is nicely captured in two heuristics:

Firm commitments are best made after the prototype works[6]

Hang on to the agony of decision as long as possible[7]

The second case, problems in determining whether or not a built system fulfills a given purpose, is mainly a problem when requirements are fundamentally immeasurable or when performance is demanded in an environment that cannot be provided for test. For example, a space system may require a failure rate so low it will never occur during any practical test (the ultraquality problem). Or, a weapon system may be required to operate in the presence of hostile countermeasures that will not exist outside a real combat environment. Neither of these requirements can be certified by test or analysis. To certify a system with requirements like these it is necessary to either substitute surrogate requirements agreed to by the client or to find alternative certification criteria.

To architect-in certifiable criteria essentially means to substitute a refined set of measurable criteria for the client's unmeasurable criteria. This requires the architect be able to convince the client of the validity of a model for connecting the refined criteria to the original criteria. One advantage of a third-party architect is the independent architect's greater credibility in making just such arguments, which may be critical to developing a certifiable system. A builder–architect, with an apparent conflict of interest, may not have the same credibility. The model that connects the surrogate criteria to the real, unmeasurable criteria may be a detailed mathematical model or may be quite heuristic. An example of the former category is failure tree analysis that tries to certify untestable reliability levels from testable subsystem reliability levels. A more heuristic model may be more appropriate for certifying performance in uncertain combat environments. While the performance required is uncertain, criteria like flexibility, reprogrammability, performance reserve, fall-back modes, and ability to withstand damage can be specified and measured.

Rational and heuristic methods can be combined to develop ultraquality systems. A good example is a paper by Jaynarayan.[8] This paper discusses the architectural principles for developing flight control computers with failure rates as low as 10^{-10} per hour. Certification of such systems is a major problem. The authors discuss a two-pronged approach. First, instead of attempting to do a failure modes and effects analysis with its enormous fault trees they design for "Byzantine failure". Byzantine failure means failure in which the failed element actively, intelligently, and malevolently attempts to cause system failure. They go on to describe formal methods for designing systems resistant to a given number of Byzantine faults, thus replacing the need to trace fault trees for each type of failure. The approach is based on designs that do not allow information or energy from a possibly failed element to propagate outside an error confinement region. The second prong is a collection of guidelines for minimizing common mode failures. These are the system failures due to design errors rather then component failures. Since one cannot design-in resistance to design failure other means are necessary. The guidelines, partially a set of heuristics, provide guidance in this otherwise nonmeasurable area.

The third and last case is uncertainty about the connection between the model and the actual system. This is an additional case where informed judgment and heuristics are needed. To reduce the uncertainty in modeling requires tests and prototypes. The best guidance on architecting prototypes is to realize that all prototypes should be purpose driven. Even when the purposes of the system are less than clear, the purposes of the prototype should be quite clear. Thus the architecting of the prototype can be approached as architecting a system, with the architect as the client.

Design progression

Progressive refinement of design is one of the most basic patterns of engineering practice. It permeates the process of architecting from models to heuristics, information acquisition, and management. Its real power, especially in systems architecting, is that it provides a way to organize the progressive transition from the ill-structured, chaotic, and heuristic processes needed at the beginning to the rigorous engineering and certification processes needed later. All can be envisioned as a stepwise reduction of abstraction, from mental concept to delivered physical system.

In software the process is known as stepwise refinement. Stepwise refinement is a specific strategy for top–down program development. The same notion applies to architecting, but is applied in-the-large to complex multidisciplinary system development. Stepwise refinement is the progressive removal of abstraction in models, evaluation criteria, and goals. It is accompanied by an increase in the specificity and volume of information recorded about the system and a flow of work from general to specialized design disciplines. Within the design disciplines the pattern repeats as disciplinary objectives and requirements are converted into the models of form of that discipline. In practice, the process is not so smooth nor continuous. It is better characterized as episodic with episodes of abstraction reduction alternating with episodes of reflection and purpose expansion.

Stepwise refinement can be thought of as a meta-process model much as the waterfall and spiral. It is not an enactable process for a specific project, but it is a model for building a project-specific process. Systems are too diverse to follow a fixed process or dogmatic formula for architecting.

Introduction by examples

Before treating the conceptually difficult process of general systems architecting, look to the roots. When a civil architect develops a building, does he or she go directly from client's words to construction drawings? Obviously not; there are many intermediate steps. The first drawings are rough floor plans showing the spatial relationships of rooms and sizes or external renderings showing the style and feel of the building. Following these are intermediate drawings giving specific dimensions and layouts. Eventually come the construction drawings with full details for the builder.

In a different domain, the beginning computer programmer is taught a similar practice. Stepwise refinement in programming means to write the central controlling routine first. Anywhere high complexity occurs, ignore it by giving it a descriptive name and making it a subroutine or function. Each subroutine or function is "stubbed", that is, given a dummy body as a placeholder. When the controlling routine is complete, it is compiled and executed as a test. Of course, it doesn't do anything useful since its subroutines are stubbed. The process is repeated recursively on each subroutine until routines can be easily coded in primitive statements in the programming language. At each intermediate step an abstracted version of the whole program exists that has the final program's structure but lacks internal details.

Both examples show progression of system representation or modeling. Progression also occurs along other dimensions. For example, both the civil architect and the programmer may create distinct alternative designs in their early stages. How are these partial designs evaluated to choose the superior approach? In the earliest stages both the programmer and the civil architect use heuristic reasoning. The civil architect can measure rough size (to estimate cost), judge the client's reaction, and ask the aesthetic opinion of others. The programmer can judge code size, heuristically evaluate the coupling and cohesion of the resulting subroutines and modules, and review functionality with the client. As their work progresses both will be able to make increasing use of rational and quantitative evaluation criteria. The civil architect will have enough details for proven cost models, the programmer can measure execution speed, compiled size, behavioral compliance, and invoke quantitative software quality metrics.

Design as the evolution of models

All architects, indeed all designers, manipulate models of the system. These models become successively less abstract as design progresses. The integrated models discussed in Chapter 9 exhibit stepwise reduction of abstraction in representation and in their design heuristics.

In Hatley-Pirbhai the reduction of abstraction is from behavioral model, to technology-specific behavioral model, to architecture model. There is also hierarchical decomposition within each component. The technology of modules is indeterminate at the top level and becomes technology specific as the hierarchy develops. The Q^2FD performance modeling technique shows stepwise refinement of customer objectives into engineering parameters. As the QFD chain continues the engineering parameters get closer to implementation until, ultimately, they may represent machine settings on the factory floor. Likewise, the structure of integrated models in software and manufacturing systems follows the same logic or progression.

Evaluation criteria and heuristic refinement

The criteria for evaluating a design progresses or evolves in the same manner as design models. In evaluation, the desirable progression is from general to system specific to quantitative. For heuristics the desirable progression is from descriptive and prescriptive qualitatives to domain-specific quantitatives and rational metrics. This progression is best illustrated by following the progression of a widely recognized heuristic into quantitative metrics within a particular discipline. Start with the partitioning heuristic:

> **In partitioning, choose the elements so that they are as independent as possible, that is, elements with low external complexity and high internal complexity.**

This heuristic is largely independent of domain. It serves as an evaluation criterion and partitioning guide whether the system is digital hardware, software, human driven, or otherwise. But the guidance is nonspecific; neither independence or complexity are defined. By moving to a more restricted domain, computer-based systems in this example, this heuristic refines into more prescriptive and specific guidelines. The literature on structured design for software (or, more generally, computer-based systems) includes several heuristics directly related to the partitioning heuristic.[9] The structure of the progression is illustrated in Figure 8.2.

1. Module fan-in should be maximized. Module fan-out should generally not exceed 7 ± 2.
2. The coupling between modules should be — in order of preference — data, data structure, control, common, and content.
3. The cohesion of the functions allocated to a particular module should be — in order of preference — functional/control, sequential, communicational, temporal, periodic, procedural, logical, and coincidental.

These heuristics give complexity and independence more specific form. As the domain restricts even farther the next step is to refine into quantitative design quality metrics. This level of refinement requires a specific domain and detailed research and is the concern of specialists in each domain. But, to finish the example, the heuristic can be formulated into a quantitative software complexity metric. A very simple example is

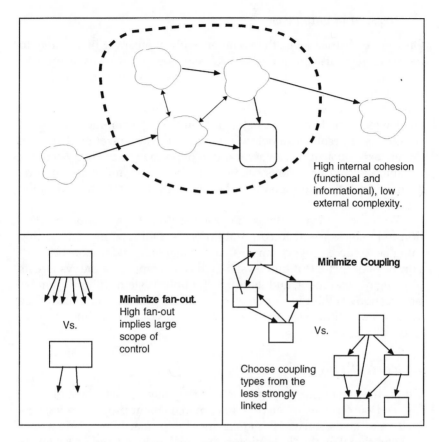

Figure 8.2   Software refinement of coupling and cohesion heuristic. The general heuristic is refined into a domain-specific set of heuristics.

> **Compute a complexity score by summing: 1 point for each line of code, 2 points for each decision point, 5 points for each external routine call, 2 points for each write to a module variable, 10 points for each write to a global variable.***

Early evaluation criteria or heuristics must be as unbounded as the system choices. As the system becomes constrained so do the evaluation criteria. What was a general heuristic judgment becomes a domain-specific guideline and, finally, a quantitative design metric.

---

* Much more sophisticated complexity metrics have been published in the software engineering literature. One of the most popular is the McCabe metric, for which there is a large automated toolset.

Progression of emphasis

On a more abstract level, the social or political meaning of a system to its developers also progresses. A system goes from being a product (something new), to a source of profit or something of value, to a policy (something of permanence). In the earliest stages of a system's life it is most likely viewed as a product. It is something new, an engineering challenge. As it becomes established and its development program progresses it becomes an object of value to the organization. Once the system exists it acquires an assumption of permanence. The system, its capabilities, and its actions become part of the organization's nature. To have and operate the system becomes a policy which defines the organization.

With commercial systems the progression is from product innovation to business profit to corporate process.[10] Groups innovate something new, successful systems become businesses, and established corporations perpetuate a supersystem which encompasses the system, its ongoing development, and its support. Public systems follow a similar progression. At their inception they are new, at their development they acquire a constituency, and they eventually become a bureaucratic producer of a commodity.

Concurrent progressions

Other concurrent progressions include risk management, cost estimating, and perceptions of success. Risk management progresses in specificity and goals. Early risk management is primarily heuristic with a mix of rational methods. As prototypes are developed and experiments conducted risk management mixes with interpretation. Solid information begins to replace engineering estimates. After system construction, risk management shifts to postincident diagnostics. System failures must be diagnosed, a process that should end in rational analysis but may have to be guided by heuristic reasoning.

Cost estimating goes through an evolution similar to other evaluation criteria. Unlike other evaluation criteria, cost is a continually evolving characteristic from the system's inception. At the very beginning the need for an estimate is highest and the information available is lowest. Little information is available because the design is incomplete and no uncertainties have been resolved. As development proceeds more information is available, both because the design and plans become more complete and because actual costs are incurred. Incurred costs are no longer estimates. When all costs are in (if such an event can actually be identified) there is no longer a need for an estimate. Cost estimating goes through a progression of declining need but of continuously increasing information.

All of the "ilities" are part of their own parallel progressions. These system characteristics are known precisely only when the system is deployed. Reliability, for example, is known exactly when measured in the field. During development reliability must be estimated from models of the system. Early in the process the customer's desires for reliability may be well known, but the reliability performance is quite uncertain. As the design progresses to lower levels the information needed to refine reliablility estimates become known — information like parts counts, temperatures, and redundancy.

Perceptions of success evolve from architect to client and back to architect. The architect's initial perception is based on system objectives determined through client interaction. The basic measure of success for the architect becomes successful certification. But once the system is delivered the client will perceive success on his or her own terms. The project may produce a system that is successfully certified but that nonetheless becomes a disappointment. Success is affected by all other conditions affecting the client at delivery and operation whether or not anticipated during design.

Episodic nature

The emphasis on progression appears to define a monotonic process. Architecting begins in judgment, rough models, and heuristics. The heuristics are refined along with the models as the system becomes bounded until rational, disciplinary engineering is reached. In practice the process is more cyclic or episodic with alternating periods of synthesis, rational analysis, and heuristic problem solving. These episodes occur during system architecting, and may appear again in later, domain-specific stages.

The occurrence of the episodes is integral to the architect's process. An architect's design role is not restricted solely to "high level" considerations. Architects dig down into specific subsystem and domain details where necessary to establish feasibility and determine client-significant performance (see Chapter 1, Figure 1.1, and the associated discussion). The overall process is one of high level structuring and synthesis (based on heuristic insight) followed by rational analysis of selected details. Facts learned from those analyses may cause reconsideration of high level synthesis decisions and spark another episode of synthesis and analysis. Eventually, there should be convergence to an architectural configuration and the driving role passes to subsystem engineers.

Architecture and design disciplines

Not very many years ago the design of a system of the complexity of several tens of thousands of logic gates was a major undertaking. It was an architectural task in the sense it was probably motivated by an explicit purpose and required the coordination of a multidisciplinary design effort. Today, components of similar and much higher complexity are the everyday building blocks of the specialized digital designer. No architectural effort is required to use such a component, or even to design a new one. In principle, art has been largely removed from the design process because the discipline has a firm scientific basis. In other words, the design discipline or domain is well worked out, and the practitioners are recognized specialists. Today it is common to discuss digital logic synthesis directly from fairly high level specifications, even if automated synthesis is not yet common practice. So there should be no surprise if systems that today tax our abilities and require architectural efforts one day become routine with recognized design methodologies taught in undergraduate courses.

The discussion of progression leads to further understanding of the distinctions between architecture and engineering. The basic distinction was reviewed in the Preface — the types of problems addressed and the tools used to address them. A refinement of the distinction was discussed in Chapter 1, the extent to which the practitioner is primarily concerned with scoping, conceptualizing, and certification. By looking at the spectrum of architecture and engineering across systems disciplines these distinctions become clearer and can be further refined. First, the methods most associated with architecting (heuristics) work best one step beyond where rational design disciplines have been worked out. This may or may not be at the forefront of component technology. Large-scale systems, by their nature, push the limits of scientific engineering at whatever level of technology development is current. But, as design and manufacturing technology change the level of integration that is considered a component, the relative working position of the architect inevitably changes. Where the science does not exist the designer must be guided by art. With familiarity and repetition, much that was done heuristically can now be done scientifically or procedurally.

However, this does not imply that where technology is mature that architecting does not exist. If it did, there would be no need for civil architects. Only systems that are relatively unique need to be architected. Development issues for unique systems contain a kernel of architectural concerns that transcend whatever technology or scientific level is current. In low technology systems, like buildings, only the nonroutine building needs to be architected. But dealing with the nonroutine, the unique, and the client/user customized is different from other engineering practices. It contains an irreducible component of art. A series of related unprecedented systems establishes a precedent. The precedent establishes recognized patterns, sometimes called architectures, of recognized worth. Further systems in the field will commonly use those established architectures, with variations more on style than in core structure.

Current development in software engineering provides an example of evolution to a design discipline. Until recently the notion of software engineering hardly existed, there was only programming. Programming is the process of assembling software from programming language statements. Programming language statements do not provide a very rich language for expressing system behaviors. They are constrained to basic arithmetic, logical, and assignment operations. To build complex system behaviors, programs are structured into higher level components that begin to express system domain concepts. But in traditional programming each of the components must be handcrafted from the raw material of programming languages.

The progression in software is through the construction and standardization of components embodying behaviors closer and closer to problem domains. Instead of programming in what was considered a "high level language", the engineer can now build a system from components close to the problem domain. The programming language is still used, but primarily to knit together prebuilt components. Programming libraries have been in common use for many years. The libraries shipped with commercial software development environments are often very large and contain extensive class or object libraries. In certain domains the gap has grown very small.

Example: The popular mathematics package MatLab allows direct manipulation of matrices and vectors. It also provides a rich library of functions targeted at control engineers, image processing specialists, and other fields. One can dispense with the matrices and vectors all together by "programming" with a graphical block diagram interface that hides the computational details and provides hundreds of prebuilt blocks. Further extensions allow the block diagram to be compiled into executable programs that run on remote machines. In fact, direct connection to physical systems for real-time control is possible.

Wherever a family of related systems is built a set of accepted models and abstractions appears and forms the basis for a specialized design discipline. If the family becomes important enough the design discipline will attract enough research attention to build scientific foundations. It will be marked as a design discipline when universities form departments devoted to it. At the same time a set of common design abstractions will be recognized as "architectures" for the family. Mary Shaw, observing the software field, finds components and patterns constrained by component and connector vocabulary, topology, and semantic constraints. These patterns can be termed "styles" of architecture in the field, as was discussed in Chapter 6.

Architecture and patterns

The progression from "inspired" architecture to formal design method is through long experience. Long experience in the discipline by its practictioners eventually yields tested patterns of function and form. Patterns, pattern languages, and styles are a formalization of this progression. Architecting in a domain matures as architects identify reusable components and repeating styles of connection. Put another way, they recognize recurring patterns of form and their relationships to patterns in problems. In a mature domain patterns in both the problem and solution domain develop rigorous expression. In digital logic (a relatively mature design domain) problems are stated in formal logic and solutions in equally mathematically well-founded components. In a less mature domain the patterns are more abstract or heuristic.

A formalization of patterns in architecture is due to Christopher Alexander.[11] Working within civil architecture and urban design, Alexander developed an approach to synthesis based on the composition of formalized patterns. A pattern is a reoccurring structure within a design domain. They consist of both a problem or functional objective for a system and solution. Patterns may be quite concrete (such as "a sunny corner") or relatively abstract (such as "masters and apprentices"). A template for defining a pattern is

1. A brief name which describes what the pattern accomplishes
2. A concise problem statement
3. A description of the problem including the motivation for the pattern and the issues in resolving the problem
4. A solution, preferably stated in the form of an instruction
5. A discussion of how the pattern relates to other patterns in the language

A pattern language is a set of patterns complete enough for design within a domain. It is a method for composing patterns to synthesize solutions to diverse objectives. In the Alexandrian method the architect consults sets of patterns and chooses from them those patterns which evoke the elements desired in a project. The patterns become the building blocks for synthesis, or suggest important elements that should be present in the building. The patterns each suggest instructions for solution structure, or contain a solution fragment. The fragments and instructions are merged to yield a system design.

Since the definitions of a pattern and a pattern language are quite general, they can be applied to other forms of architecture. The ideas of patterns and pattern languages are now a subject of active interest in software engineering.* Software architects often use the term "style"

---

* A brief summary with some further references is Bercuzk, C., Hot topics, finding solutions through pattern languages, *IEEE Computer*, p. 75, 1995.

to refer to reoccurring patterns in high level software design. Various authors have suggested patterns in software using a pattern template similar to that of Alexander. An example of a software pattern is "callbacks and handlers", a commonly used style of organizing system-dependent bindings of code to fixed behavioral requirements.

The concept of a style is related to Alexandrian patterns since each style can be described using the pattern template. Patterns are also a special class of heuristic. A pattern is a prescriptive heuristic describing particular choices of form and their relationship to particular problems. Unlike patterns, heuristics are not tied to a particular domain.

Although the boundaries are not sharp, heuristics, patterns, styles, and integrated design methods can be thought to form a progression. Heuristics are the most general, spanning domains and categories of guidance. However, they are also the least precise and give the least guidance to the novice. Patterns are specially documented, prescriptive heuristics of form. They prescribe (perhaps suggest) particular solutions to particular problems within a domain. A style is still more precisely defined guidance, this time in the form of a domain-specific structure. Still farther along the maturity curve are fully integrated design methods. These have domain-specific models and specific procedures for building the models, transforming models of one type into another type, and implementing a system from the models.

Thus the largest-scale progression is from architecting to a rigorous and disciplined design method, one that is essential to the pronouncement theory of design. Along the way the domain acquires heuristics, patterns, and styles of proven worth. As the heuristics, patterns, and styles become more specific, precise, and prescriptive they give the most guidance to the novice and come closest to the normative (what *should* be) theory of design. As design methods become more precise and rigorous they also become more amenable to scientific study and improvement. Thus the progression carries from a period requiring (and encouraging) highly creative and innovative architecting to one defined by quantifiable and provable science.

Civil architecture experience suggests that at the end of the road there will still be a segment of practice that is best addressed through a fusion of art and science. This segment will be primarily concerned with the client's view of a system, and will seek to reconcile client satisfaction and technical feasibility. The choice of method will depend on the question. If you want to know how a building will fare in a hurricane, you know to ask a structural engineer. If you want the building to express your desires, and do so in a way beyond a rote calculation of floor space and room types, you know to ask an architect.

Conclusions

A fundamental challenge in defining a system architecting method or
a system architecting tool kit is its unstructured and eclectic nature.
Architecting is synthesis oriented, and operates in domains and with
concerns that preclude rote synthesis. Successful architects proceed
through a mixture of heuristic and rational or scientific methods. One
meta-method that helps organize the architecting process is that of
progression.

Architecting proceeds from the abstract and general to the domain
specific. The transition from the unstructured and broad concerns of
architecting to the structured and narrow concerns of developed design
domains is not sharp. It is progressive as abstract models are gradually
given form through transformation to increasingly domain-specific
models. At the same time all other aspects of the system undergo
concurrent progressions from general to specific.

While the emphasis has been on the heuristic and unstructured
components of the process, that is not to undervalue the quantitative
and scientific elements required. The rational and scientific elements
are tied to the specific domains where systems are sufficiently constrained to allow scientific study. The broad outlines of architecting are
best seen apart from any specific domain. A few examples of the intermediate steps in progression were given in this chapter. The next chapter brings these threads together by showing specific examples of models and their association with heuristic progression. Some of the
examples in Chapter 9 assemble these threads for the domains discussed in Part 2. In addition, there are examples for other major
domains not discussed in Part 2.

Exercises

1. Find an additional heuristic progression by working from the
   specific to the general. Find one or more related design heuristics
   in a technology-specific domain. Generalize those heuristics to
   one or more heuristics that apply across several domains.

2. Find an additional heuristic progression by working from the
   general to the specific. Choose one or more heuristics from Appendix A. Find or deduce domain specific heuristic design
   guidelines in a technology domain familiar to you.

3. Examine the hypothesis that there is an identifiable set of "architectural" concerns in a domain familiar to you. What issues
   in the domain are unlikely to be reducible to normative rules or
   rational synthesis?

4. Trace the progression of behavioral modeling throughout the development cycle of a system familiar to you.

5. Trace the progression of physical modeling throughout the development cycle of a system familiar to you.

6. Trace the progression of performance modeling throughout the development cycle of a system familiar to you.

7. Trace the progression of cost estimation throughout the development cycle of a system familiar to you.

Notes and References

1. Rechtin, E., *Systems Architecting, Creating & Building Complex Systems*, Prentice-Hall, Englewood Cliffs, NJ, 1991, 14.
2. ADARTS Guidebook, p. 8.
3. ADARTS Guidebook, p. 8.
4. Rechtin, E., *Systems Architecting, Creating & Building Complex Systems*, Prentice-Hall, Englewood Cliffs, NJ, 1991, 312.
5. Selby, R. W. and Basili, V. R., Analyzing error-prone system structure, *IEEE Trans. Software Eng.*, SE-17(2), 141, February 1991.
6. Comments to the authors on Rechtin 1991 by L. Bernstein in the context of large network software systems.
7. Rechtin, E., *Systems Architecting, Creating & Building Complex Systems*, Prentice-Hall, Englewood Cliffs, NJ, 1991, 315.
8. Jaynarayan, H. and Harper, R., Architectural principles for safety-critical real-time applications, *Proc. IEEE*, 82(1), 25, January 1994.
9. Yourdon, E. and Constantine, L. L., *Structured Design: Fundamentals of a Discipline of Computer Program and Systems Design*, Yourdon Press, Englewood Cliffs, NJ, 1979.
10. Baumeister, B., personal communication.
11. Alexander, C., *The Timeless Way of Building*, Oxford University Press, New York; and *A Pattern Language: Towns, Buildings, Construction*, Oxford University Press, New York, 1977.

chapter nine

Integrated modeling methodologies

Introduction

The previous two chapters explored the model views, model integration, and progression along parallel paths. This chapter brings together the threads of Part 3 by presenting examples of integrated modeling methodologies. The methodologies are divided by domain specificity, with the first models more nearly domain independent and later models more domain specific.

Architecting clearly is domain dependent. A good architect of avionics systems, for example, may not be able to effectively architect social systems. Hence, there is no attempt to introduce a single set of models suitable for architecting everything. The models of greatest interest are those tied to the domain of interest, although they must support the level of abstraction needed in architecting. The integrated models chosen for this chapter include two principally intended for real-time, computer-based, mixed hardware/software systems (H/P and Q²FD), three methods for software-based systems, one method for manufacturing systems, and, conceptually at least, some methods for including human behavior in sociotechnical system descriptions.

The examples for each method were chosen from relatively simple systems. They are intended as illustrations of the methods and their relevance to architectural modeling and to fit within the scope of the book. They are not intended as case studies in system architecting.

General integrated models

The two most general of the integrated modeling methods are Hatley/Pirbhai (H/P) and Q²FD.

Hatley/Pirbhai — computer-based reactive systems

A computer based reactive system is a system which senses and reacts to events in the physical world, and in which much of the implementation complexity is in programmable computers. Multifunction automobile engine controllers, programmable manufacturing robots, and military avionics systems (among many others) all fall into this category. They are distinguished by mixing continuous, discrete, and discrete event logics and being implemented largely through modern computer technology. The integrated models used to describe these systems emphasize detailed behavior descriptions, physical descriptions matched to software and computer hardware technologies, and some performance modeling. Recent efforts at defining an engineering of computer-based systems discipline[1] are directed at systems of this type.

Several different investigators have worked to build integrated models for computer-based reactive systems. The most complete example of such an integration is the Hatley/Pirbhai (H/P) methodology.* Other methods, notably, the FFBD method used in computer tools by Ascent Logic and the StateMate formalism, are close in level of completeness. This section concentrates on the structure of H/P. With the concepts of H/P in mind it is straightforward to make a comparative assessment of other tools and methods.

H/P defines a system through three primary models: two behavioral models (the "requirements model" [RM] and the "enhanced requirements model" [ERM]) and a model of form called the "architecture model" (AM). The two behavioral models are linked through an embedding process. The behavioral and form models are linked by static allocation tables. The performance view is linked statically through timing allocation tables. More complex performance models have been integrated with H/P, but descriptions have only recently been published. The data view is represented by a dictionary. This dictionary provides a hierarchical data element decomposition, but does not provide a syntax for defining dynamic data relationships. No managerial view is provided, although managerial metrics have been defined for models of the H/P type.

Both behavioral models are based on DeMarco-style data flow diagrams. The data flow diagrams are extended to include finite state and event processing through what is called the "control model". The control model uses data flow diagram syntax with discrete events and

---

* Wood, D. P. and Wood, W. G., Comparative Evaluation of Four Specification Methods for Real-Time Systems, Software Engineering Institute Technical Report, CMU/SEI-89-TR-36, 1989. This study compared four popular system modeling methods. Their conclusion was that the Hatley/Pirbhai method was the most complete of the four, though similarities were more important than the differences. In the intervening time many of the popular methods have been extended and additional tools reflecting multiview integration have begun to appear.

finite-state machine processing specifications. The behavioral modeling syntax is deliberately nonrigorous and is not designed for automated execution. This "lack" of rigor is deliberate; it is intended to encourage flexibility in client/user communication. The ERM is a superset of the requirements model. It surrounds the core behavioral model and provides a behavioral specification of the processing necessary to resolve the physical interfaces into problem domain logical interfaces. The ERM defines implementation-dependent behaviors, such as user interface and physical I/O.

Example: microsatellite imaging system

Some portions of a H/P model formulated for the imaging (camera) subsystem of a developmental microsatellite provide an illustration of the H/P concepts. This example is to present the flavor of the H/P idiom for architects, not to fully define the imaging system. The level chosen is representative of that of a subsystem architecture (not all architecting has to be done on systems of enormous scale!). Figure 9.1 shows the top level behavioral model of the imaging system, defined as a data flow diagram (DFD). Each circle on the diagram represents a data-triggered function or process. So, for example, process number 2, "Evaluate Image", is triggered by the presence of a "Raw Image" data element. Also from the diagram, process number 2 produces a data element of the same type (the outgoing arrow labeled "Raw Image") and another data element called "Image Evals".

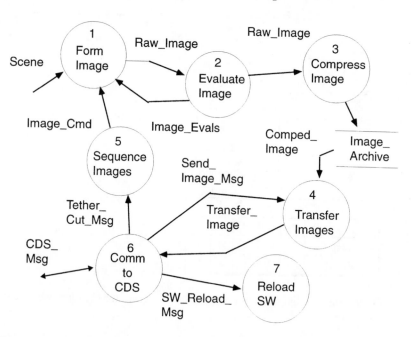

Figure 9.1 Top level data flow diagram for a microsatellite imaging system.

Each process in the behavior model is defined either by its own DFD or by a textual specification. During early development processes may be defined with brief and nonrigorous textual specifications. Later, as processes are allocated to physical modules, the specifications are expanded in greater detail until implementation-appropriate rigor is reached. Complex processes may have more detailed specifications even early in the process. For example, in Figure 9.2 process number 1, "Form Image", is expanded into its own diagram.

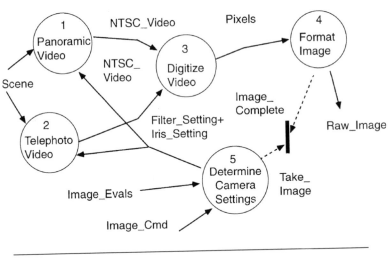

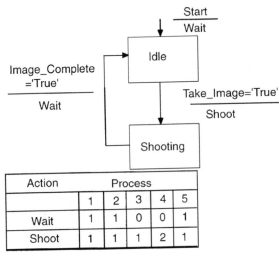

Figure 9.2 Expanded data and control flow diagram for the top level process "Form Image".

Figure 9.2 also introduces control flow. The dotted arrows indicate flow of control elements, and the solid line into which they flow is a control specification. The control specification is shown as part of the same figure. Control flows may be interpreted either as continuous time, discrete valued data items, or as discrete events. The latter interpretation is more widely used, although it is not preferred in the published H/P examples. The control specification is a finite-state machine, here shown as a state transition diagram, although other forms are also possible. The actions produced by the state machine are to activate or deactivate processes on the associated DFD.

All data elements appearing on a diagram are defined in the data dictionary. Each may be defined in terms of lower level data elements. For example, the flow "Raw Image" appearing in Figure 9.2 appears in the data dictionary as:

$$\text{Raw Image} = 768\{484\{\text{Pixel}\}\}$$

indicating, in this case, that Raw Image is composed of 768×484 repetitions of the element "Pixel". At early stages Pixel is defined qualitatively as a range of luminance values. In later design stages the definition will be augmented, though not replaced, by a definition in terms of implementation-specific data elements.

In addition to the two behavior models the H/P method contains an "architecture model". The architecture model is the model of form that defines the physical implementation. The architecture model is hierarchical. It allows sequential definition in greater detail by expansion of modules. Figure 9.3 shows the paired architecture flow and interconnect models for the microsatellite imaging system.

The H/P block diagram syntax partitions a system into modules, which are defined as physically identifiable implementation elements. The flow diagram shows the exchange of data elements among the modules. Which data elements are exchanged among the modules is defined by the allocation of behavioral model processes to the modules.

The interconnection model defines the physical channels through which the data elements flow. Each interconnect is further defined in a separate specification. For example, the interconnect "Tputer Channel 1" connects the processor module and the camera control module. Allocation requires camera commands to flow over the channel. Augmentations to the data dictionary define a mapping between the logical camera commands and the line codes of the channel. If the channel requires message framing, protocol processing, or the like, it is defined in the interconnection specification. Again, the level of detail provided can vary during design, based on the interface's impact on risk and feasibility.

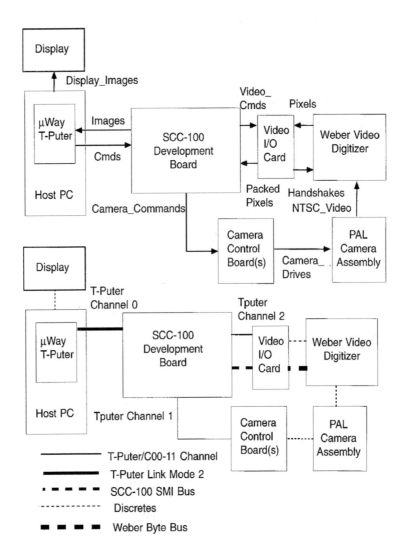

Figure 9.3 Architecture flow and interconnect diagrams for the example microstaellite imaging system laboratory prototype.

Quantitative QFD (Q²FD) — performance-driven systems

Many systems are driven by quantitatively stated performance objectives. These systems may also contain complex behavior or other attributes, but its performance objectives are of most importance to the client. For these systems it is common practice to take a performance-centered approach to system specification, decomposition, and synthesis. A particularly attractive way of organizing such a decomposition is through extended quality function deployment (QFD) matrices.[2]

QFD is a Japanese-originated method for visually organizing the decomposition of customer objectives.[3] It builds a graphical hierarchy of how customer objectives are addressed throughout a system design, and carries the relevance of customer objectives throughout design. A Q^2FD-based approach requires that the architect:

1. Identify a set of performance objectives of interest to the customer. Determine appropriate values or ranges for meeting these objectives through competitive analysis.
2. Identify the set of system level design parameters that determine the performance for each objective. Determine suitable satisfaction models that relate the parameters and objectives.
3. Determine the relationships of the parameters and objectives and the interrelationships among the parameters. Which affect which?
4. Set one or more values for each parameter. Multiple values may be set, for example, minimum, nominal, and target. Additional slots provide tracking from detailed design activities.
5. Repeat the process iteratively using the system design parameters as objectives. At each stage the parameters at the next level up become the objectives at the next level down.
6. Continue the process of decomposition at as many levels as desired. As detailed designs are developed their parameter values can flow up the hierarchy to track estimated performance for customer objectives. The structure is illustrated in Figure 9.4.

Unfortunately, QFD models for real problems tend to produce quite large matrices. Since they can map directly to computer spreadsheets this causes no difficulty in modern work environments, but it does cause a problem in presenting an example. Also, the graphic of the matrix shows the result, but hides the satisfaction models. The satisfaction models are equations, simulations, or assessment processes necessary to determine the performance measure value. The original reference on QFD by Hauser contains a qualitative example of using QFD for objective decomposition, as do other books on QFD. Two papers by one of the present authors[4] contain detailed, quantitative examples of QFD performance decomposition using analytical engineering models.

Integrated modeling and software

Chapters 7 and 8 introduced the ideas of model views and stepwise refinement-in-the-large. Both of these ideas have featured prominently in the software engineering literature. Software methods have been the principal sources for detailed methods for expressing multiple views and development through refinement. Software engineers have developed

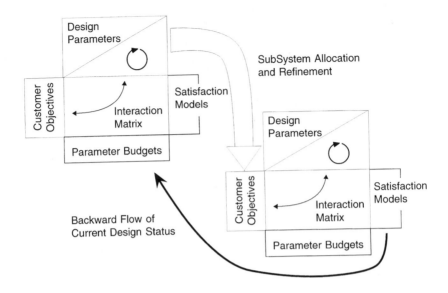

Figure 9.4 QFD hierarchy tree. The basic matrix shows the interrelationships of customer objectives and engineering design parameters. QFD matrices are arranged in a hierarchy that can mirror the decomposition of a system into subsystems and modules.

several integrated modeling and development methodologies that integrate across views and employ explicit heuristics. Three of those methods are described in detail: structured analysis and design, ADARTS, and OMT.

The three methods are targeted at different kinds of software systems. Structured analysis and design was developed in the late 1970s and early 1980s and is intended for single-threaded software systems written in structured procedural languages. ADARTS is intended for large, real-time, multithreaded systems written in Ada. Its methods do generalize to other environments. OMT is intended for data base-intensive systems, especially those written in object-oriented programming languages.

Structured analysis and design

The first of the integrated models for software was the combination of structured analysis with structured design.[5] The software modeling and design paradigms established in that book have continued to the present as one of the fundamental approaches to software development. Structured analysis and design models two system views, uses a variety of heuristics to form each view, and connects to the management view through measurable characteristics of the analysis and design models (metrics).

The method prescribes development in three basic steps. Each step is potentially quite complex and is composed of many internal steps of refinement. The first step is to prepare a data flow decomposition of the system to be built. The second step is to transform that data flow decomposition into a function and module hierarchy that fully defines the structure of the software in subroutines and their interaction. The design hierarchy is then coded in the programming language of choice. The design hierarchy can be mechanically converted to software code (several tools do automatic forward and backward conversion of structured design diagrams and code). The internals of each routine are coded from the included process specifications, though this requires human effort.

The first step, known as structured analysis, is to prepare a data flow decomposition of the system to be built. A data flow decomposition is a tree hierarchy of data flow diagrams, textual specifications for the leaf nodes of the hierarchy, and an associated data dictionary. This method was first popularized by DeMarco,[6] though the ideas had appeared previously and it has since been extensively modified and represented. Figures 9.1 and 9.2, discussed in the previous example, are examples of data flow diagrams. Behavioral analysis by data flow diagram originated in software and has since been applied to more general systems as well. The basic tenets of structured analysis are

1. Show the structure of the problem graphically, engaging the mind's ability to perceive structure and relationships in graphics.
2. Limit the scope of information presented in any diagram to five to nine processes and their associated data flows.
3. Use short (<1 page) free-form and textual specifications at the leaf nodes to express detailed processing requirements.
4. Structure the models so each piece of information is defined in one and only one place. This eases maintenance.
5. Build models in which the processes are loosely coupled, strongly cohesive, and which obey a defined syntax for balance and correctness.

Structured design follows structured analysis and transforms a structured analysis model into the framework for a software implementation. The basic structured design model is the structure chart. A structure chart, as illustrated in Figure 9.5, shows a tree hierarchy of software routines. The arrows connecting boxes indicate the invocation of one routine or subroutine by another. The circles, arrows, and names show the exchange of variables and are known as data couples. Additional symbols are available for pathological connection among routines, such as unconditional jumps. Each box on the structure chart is

linked to a textual specification of the requirements for that routine.
The data couples are linked to a data dictionary.

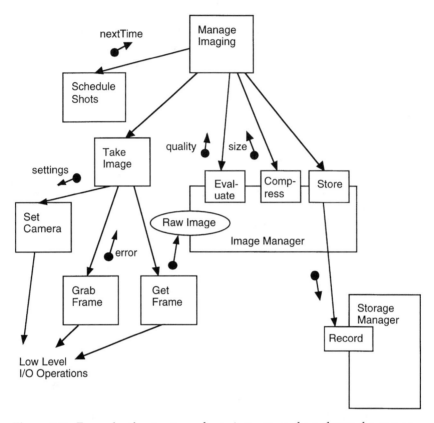

Figure 9.5 Example of a structure chart. A structure chart shows the components and interfaces for a hierarchically structured software system. This simple
example illustrates routines as components (boxes), collection of routines into
modules (boxes encapsulating other boxes), invocation of one routine by another (arrows connecting boxes), data elements (the labeled oval), and data
passing among routines (solid circles with arrows).

Structure charts are closely aligned with the ideas and methods of
structured programming, which was a major innovation at the time
structured design was introduced. Structure charts can be mechanically
converted to nested subroutines in languages which support the structured programming concepts. In combination, the chart structure, the
interfaces shown on the chart, and the linked module specifications
define a compilable shell for the program and an extended set of code
comments. If the module specifications are written formally, they can
be the module's program design language or can be compiled as module precondition and postcondition assertions.

The structured analysis and design method goes farther in providing detailed heuristics for transformation of an analysis model into a structure chart and for evaluation of alternative designs. The heuristics are strong, in the sense that they are stated procedurally. However, they are still heuristics because their guidance is provisional and subject to interpretation in the overall context of the problem. The transformation is a type of refinement or reduction of abstraction. The data flow network of the analysis phase defines data exchange, but it does not define execution order beyond that implied by the data flow. Hence the structure chart removes the abstraction of flow of control by fixing the invocation hierarchy. The heuristics provided are of two types. One type gives guidelines for transforming a data flow model fragment into a module hierarchy. The other type measures comparative design quality to assist in selection among alternative designs. Examples of the first type include:

Step one: Classify each data flow diagram as "transform oriented" or "transaction oriented" (these terms are further defined in the method).

Step two: In each case, find either the "transform center" or the "transaction center" of the diagram and begin factoring the modules from there.

Further heuristics follow for structuring transform-centered and transaction-centered processes. In the second category are several quite famous design heuristics:

Choose designs which are loosely coupled. Coupling, from loosest to tightest, is measured as: data, data structure, control, global, and content.

Choose designs in which the modules are strongly cohesive. Cohesion is rated as (from strongest to weakest): functional, sequential, communicational, procedural, temporal, logical, and coincidental.

Choose modules with high fan-in and low fan-out.

As was discussed in Chapter 8, very general and domain-specific heuristics may be related by chains of refinement. In structured analysis and design the software designer transforms rough ideas into DFDs, DFDs into structure charts, and structure charts into code. At the same time heuristic guidelines like "strive for loose coupling" are given measurable form as the design is refined into specific programming constructs.

Various efforts have also been made to tie structured analysis and design to managerial models by predicting cost, effort, and quality from measurable attributes of DFDs or structure charts. This is done both directly and indirectly. A direct approach computes a system complexity metric from the DFDs or the structure charts. That complexity metric then must be correlated to effort, cost, schedule, or other quantities of management interest. A later work by DeMarco[7] describes a detailed approach on these lines, but the suggested metrics have not become popular nor have they been widely validated on significant projects. Other metrics, such as function or feature points, that are more loosely related to structured analysis decompositions have found some popularity. Software metrics is an ongoing research area and there is a growing body of literature on measurements that appear to correlate well with project performance.

An alternative linkage is indirect by using the analysis and design models to guide estimates of the most widely accepted metrics, the constructive cost model (COCOMO) and effective lines of code (ELOC). COCOMO is Barry Boehm's famous effort estimation formula. The model predicts development effort from a formula involving the total lines of code, an exponent dependent on the project type, and various weighting factors. One problem with the original COCOMO model is that it does not differentiate between newly written lines of code and reused code. One method (there are others) of extending the COCOMO model is to use ELOC in place of total lines of code. ELOC measures the size of software project giving allowance for modified and reused code. A new line of code counts for one ELOC; modified and unmodified, reused code packages count for somewhat less. The weight factors given to each are typically determined organization-by-organization based on past measurements. The counts by subtype are summed with their weights and the total treated as new lines in the COCOMO model.

The alternative approach is to use the models to guide ELOC estimation. Early in the process, when no code has been written, the main source of error in COCOMO is likely to be errors in the ELOC estimate. With a data flow model in hand, engineers and managers can go through it process-by-process and compare the requirements to past efforts by the organization. This, at least, structures the estimation problem to identifiable pieces. Similarly, the structured design model can be used in the same way, with estimates of the ELOC for each module flowing upward into a system level estimate. As code is written the estimates become facts and, hopefully, the estimated and actual efforts will converge. Of course, if the organization is incapable of producing a given ELOC level predictably, any linkage of analysis and design models to managerial models is moot.

The architect needs to be cognizant of these issues insofar as they affect judgments of feasibility. As the architect develops models of the system they should be used jointly by client and builder. As the client's value judgments should be made in the context of the models, the builder's estimates should be as well. If builder organizations have a lot of variance in what effort is required to deliver a fixed complexity system, then that variance is a risk to the client.

ADARTS

Ada-Based Design Approach for Real Time Systems (ADARTS) is an extensively documented example of a more advanced integrated modeling method for software. The original work on data flow techniques was directly tied to the advanced implementation paradigms of the day. In a similar way, the discrete event system-oriented specification methods like H/P can be closely tied to implementation models. In the case of real-time, event-driven software one of the most extensive methods is the ADARTS[8] methodology of the Software Productivity Consortium. The ADARTS method combines a discrete event-based behavioral model with a detailed, stepwise refined, physical design model. The behavioral model is based on DFDs extended with the discrete event formalisms of Ward and Mellor[9] (which are similar to those of H/P). The physical model includes evolving abstractions for software tasks or threads, objects, routines, and interfaces. It also includes provisions for software distributed across separate machines and their communication. ADARTS includes a catalog of heuristics for choosing and refining the physical structure through several levels of abstraction.

ADARTS links the behavioral and physical models through allocation tables. Performance decomposition and modeling are considered specifically, but only in the context of timing. There are links to sophisticated scheduling formalisms and SPC developed simulation methodologies as part of this performance link. Again, managerial views are supported through metrics, where they can be calculated from the models. Software domain-specific methods can more easily perform the management metric integration since a variety of cost and quality metrics that can be (at least roughly) calculated from software design models are known.

The example shown is a simplified version of the first two design refinements required by ADARTS applied to the microsatellite imaging system originally discussed in the Hatley/Pirbhai example. The resulting diagrams are shown in Figure 9.6. The ADARTS process takes the functional hierarchy of the behavioral model and breaks it into undifferentiated components. Each component is shown on the diagram by a cloud-shaped symbol, indicating its specific implementation structure has not yet been decided. The clouds exchange data elements dependent on the behavior allocated to each cloud. Various heuristics and engineering judgment guide the choice of clouds.

The next refinement specializes the clouds to tasks, modules or objects, and routines. ADARTS actually uses several discrete steps for this, but they are combined into one for the simple example given here. Again, the designer uses ADARTS provided heuristics and individual

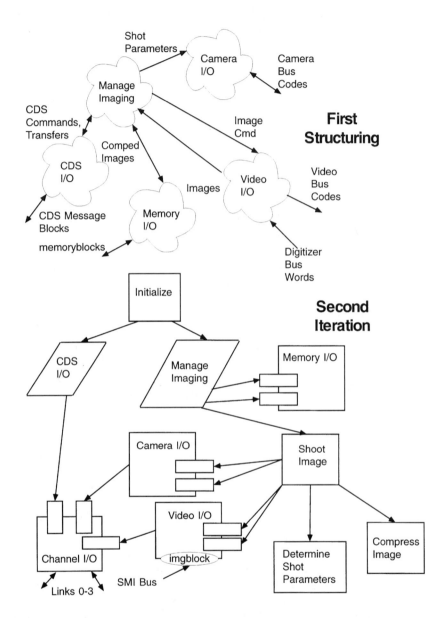

Figure 9.6 Example of a two-step design refinement in ADARTS.

judgment in making the refinements. In the example the two tasks result from the need to provide asynchronous external communications and overall system control. The clouds which hide the physical and logical interfaces to hardware are multi-entry modules. The entries are chosen from the principal user functions addressed by the interface. For example, the Camera I/O module has entries that correspond to its controls (camera shutter speed, camera gain, filter wheel position, etc.). The single thread sequence of taking an image is implemented as a simple routine calling tree.

To avoid diagram clutter, the diagram is not fully annotated with the data elements and their flow directions. In complex systems diagram clutter is a serious problem, and one not well addressed by existing tools. The architect needs to suppress some detail to process the larger picture. But correct software ultimately depends on getting each detail right. In the second part of the figure the arrowed lines indicate direction of control, not direction of data flow. Additional enhancements specify flow. The next step in the ADARTS process, not shown here, is to refine the task and module definitions once again into language- and system-specific software units. ADARTS as published assumes the use of the Ada language for implementation. When implementing in the Ada language, tasks become Ada tasks and multi-entry modules become packages. The public/private interface structure of the modules is implemented directly using constructs of the Ada language. Other languages can be accommodated in the same framework by working out language and operation-specific constructs equivalent to tasks, modules, and routines. For example, in the C language there is no language construct for tasks or multi-entry modules. But multi-entry modules can be implemented in a nearly standard way using separately compilable files on the development system, the static declaration, and suitable header files. Similarly, many implementation environments support multitasking, and some development environments supply task abstractions for the programmer's use.

Once again, the pattern of stepwise reduction of abstraction is evident. Design is conducted through steps, and at each step a model of the client needs is refined in an implementation environment-dependent way. In environments well matched to the problem-modeling method the number of steps is small; client-relevant models can be nearly directly implemented. In less well-suited environments layers of implementation abstraction become necessary.

OMT

The Hatley/Pirbhai method and its cousins are derived from structured functional decomposition, structured software design, and hardware system engineering practice. The object-oriented methods, of which OMT[10] is a leading member, derive from data-oriented and relational data base software design practice. Relational modeling methods focus solely on data structure and content and are largely restricted to data base design (where they are very powerful). Object-oriented methods package data and functional decomposition together. Where structured methods build a functional decomposition backbone on which they attempt to integrate a data decomposition, the object-oriented methods emphasize a data decomposition on which the functional decomposition is arranged. Some problems naturally decompose nicely in one method and not in the other. Complex systems can be decomposed with either, but either approach will yield subsections where the dominant decomposition paradigm is awkward.

OMT combines the data (relational), behavioral, and physical views. The physical view is well captured for software-only systems, but specific abstractions are not given for hardware components. While, in principle, OMT and other object-oriented methods can be extended to mixed hardware/software systems and even more general systems, there is a lack of real examples to demonstrate feasibility. Broad, real experience has been obtained only for predominantly software-based systems.

Neither the OMT or other object-oriented methods substantially integrate the performance view. Again, managerial views can be integrated to the extent that useful management metrics can be derived from the object models. Because of the software orientation of object-oriented methods there have been some efforts to integrate formal methods into object models.

As an example of the key ideas of object-oriented methods we present part of an object model. Object modeling starts by identifying classes. Classes can be thought of (for those unfamiliar with object concepts) as templates for objects or types for abstract data types. They define the object in terms of associated simple data items and functions associated with the object. Classes can specialize into subclasses which share the behavior and data of their parent while adding new attributes and behavior. Objects may be composed of complex elements or relate to other objects. Both composition or aggregation and association are part of a class system definition. The microsatellite imager described in the preceding section will produce images of various types. Consider an image data base for storing the data produced by the imager. A basic class diagram is shown in Figure 9.7 to illustrate specific instances of some of the concepts.

A core assumption, which the model must capture, is that images are of several distinct but related types. The actual images captured by the cameras are single gray-scale images. Varying sets of gray-scale images captured through different filters are combined into composite multiband images, with a particular gray-scale image possibly part of several composite images. In addition, images will be displayed on multiple platforms so we demand a common "rendered" image form. Each of these considerations is illustrated in Figure 9-7.

The top box labeled "Image" indicates there is a data class "Image". That class contains two data attributes — CompressedSize and ExpandedSize — and three operations or "methods" (the functions Render(), Compress(), and Expand()). The triangle-boxed lines down to the class boxes "MultiBand Image" and "Single Image"

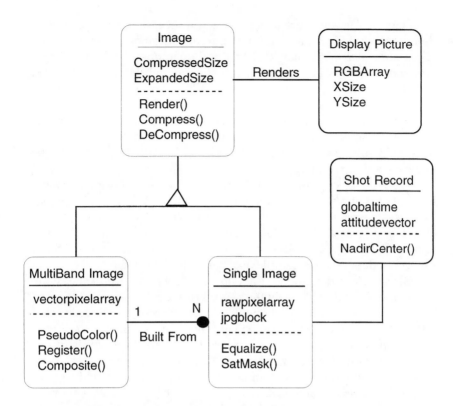

Figure 9.7 Class structure diagram for an image data base.

define those two classes as subclasses of Image. As subclasses they are different than their parent class, but inherit the parent class's data attributes and associated methods.

The class Single Image is the basic image data object descriptor. It contains two data arrays, one to hold the raw image and the other to hold the compressed form. It also has basic image processing methods associated. A multiband image is actually made up of many single images suitably processed and composited. This is defined on the diagram by the round headed line connecting the two class boxes. The labeling defines a 1- to N-way association named "Built From". The additional methods associated with MultiBand Image build the image from its associated simple images.

The two additional associations define other record keeping and display. The associated line between Single Image and Shot Record associates an image with a potentially complicated data record of when it was taken and the conditions at that moment. The association line to Display Picture shows the association of an image with a common display data structure. Both associations, in these cases, are one to one.

Figure 9.7 is considerably simplified on several points. A complete definition in OMT would require various enhancements to show actual types associated with data attributes and operations. In addition, several enhancements are required to distinguish abstract methods and derived attributes. A brief explanation of the former is in order. Consider the method Compress in the class Image. The implementation of image compression may be quite different for a single gray-scale image and for a composited multiband image. A method that is re-implemented in subclasses is called either virtual or abstracted and may be noted by a diagrammatic enhancement.

The logic of object-oriented methods is to decompose the system in a data-first fashion, with functions and data tightly bound together in classes. Instead of a functional decomposition hierarchy we have a class hierarchy. Functional definition is deferred to the detailed definition of the classes. The object-oriented logic works well where data and especially data relation complexity dominate the system.

Object-oriented methods also follow a stepwise reduction of abstraction approach to design. From the basic class model, we next add implementation-specific considerations. These will determine whether or not additional model refinements or enhancements are required. If the implementation environment is strongly object oriented there will be direct implementations for all of the model constructs. For example, in an object-oriented data base system one can declare a class with attributes and methods directly and have long-term storage (or "persistence") automatically managed. In nonobject environments it may be necessary to manually flatten class hierarchies and add manual implementations of the model features. Manual adjustments can be captured in an intermediate model of similar type. The steps of abstraction reduction depend on the environment. In a favorable implementation environment the model nearest to the client's domain can be implemented almost directly. In unfavorable environments we have no choice but to add additional layers of refinement.

Performance integration: scheduling

One area of nonfunctional performance which is very important to software, and for which there is a large body of science, is timing and scheduling. Real-time systems must perform their behaviors within a specified timeline. Absolute deadlines produce "hard real-time systems". More flexible deadlines produce "soft real-time systems". The question of whether or not a given software design will meet a set of deadlines has been extensively studied.* To integrate these timing

---

* Stankovic, J. A., Spuri, M., Di Natale, M., and Buttazzo, G. C., Implications of classical scheduling results for real-time systems, *IEEE Computer*, p. 16, June 1995, provides a good tutorial introduction to the basic results and a guide to the literature.

considerations with the design requires integration of scheduling and scheduling analysis.

In spite of the extensive study, scheduling design is still at least partly art. Theoretical results yield scheduling and performance bounds, and associated scheduling rules, but can do so only for relatively simple systems. When system functions execute interchangeably on parallel processors, run times are random, and events requiring reaction occur randomly, there are no deducible, provably optimal solutions. Some measure of insight and heuristic guidance is needed to make the system both efficient and robust.

Integrated models for manufacturing systems

The domain of manufacturing systems contains nice examples of integrated models. The modeling method of Baudin[11] integrates four modeling components (data flow, data structure, physical manufacturing flow, and cash flow) into an interconnected model of the manufacturing process. Baudin further shows how this model can then be used to analyze production scheduling under different algorithms. The four parts of the core model are

1. A data flow model using the notations of DeMarco and state transition models
2. A data model based on entity-relationship diagrams
3. A material flow model of the actual production process — the model of physical form — using ASME and Japanese notations
4. A funds flow model

These parts, which mostly use the same component models familiar from previous discussion, form an integrated architect's toolkit for the manufacturing domain. They are shown in Figure 9.8. The data flow models are in the same fashion as the requirements model of Hatley/Pirbhai. The data model is more complex and uses basic object-oriented concepts. In the material flow model the progression of removal of abstraction is taken to a logical conclusion. Since the physical architecture of manufacturing systems is restricted, the architecture model components are similarly restricted. Baudin incorporates, in fact, exploits, the restricted physical structure of manufacturing systems by using a standardized notation for the physical or form model.

Baudin further integrates domain-specific performance and system models by considering the relationship to production planning in its several forms (MRP-II, OPT, JIT). As he shows, these formalisms can be usefully placed into context on the integrated models. In the terms used in Chapter 7, this is a form of performance model integration.

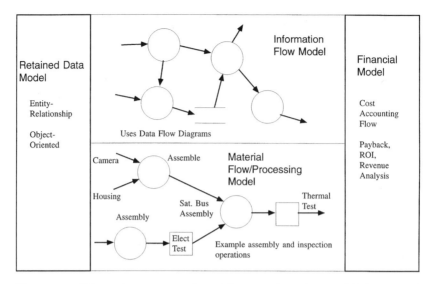

Figure 9.8 Schematic representation of the manufacturing model elements of the method of Baudin. The information flow and retained data models are the same as before. The material flow and processing model uses manufacturing specific symbology for assembly, inspection, and other operations.

Integrated models for sociotechnical systems

On the surface, the modeling of sociotechnical systems is not greatly different from other systems, but the deeper reality is quite different. The physical structure of sociotechnical systems is the same as of other systems, though it spans a considerable range of abstraction, from the concrete and steel of transportation networks to the pure laws and policy of communication standards. But people and their behavior are inextricably part of sociotechnical systems. Sociotechnical system models must deal with the wide diversity of views and the tension between facts and perceptions, as surely as they must deal with the physics of the physical systems.

Physical system representation is the same as in other domains. A civil transport system is modeled with transportation tools. A communications network is modeled by communications tools. If part of the system is an abstract set of laws or policies it can be modeled as proposed laws and policies. The fact that part of the system is abstract does not prevent its representation, but it does make understanding the interaction between the representation and the surrounding environment difficult. In general, modeling the physical component of sociotechnical systems does not present any insurmountable intellectual challenges. The unique complexity is in the interface to the humans who are components of the system and in their joint behavior.

In purely technical systems the environment and the system interact. But it is uncommon to ascribe intelligent, much less purposively hostile behavior to their environments.* However, human systems constantly adapt. If an intelligent transport system unclogs highways, people may move farther away from work and reclog the highways until a new equilibrium is reached. A complete model of sociotechnical system behavior must include the joint modeling of system and user behavior, including adaptive behavior on the part of the users.

This joint behavioral modeling is one area where modeling tools are lacking. The tools that are available fall into a few categories: econometrics, experimental microeconomics and equilibrium theory, law and economics, and general system dynamics. Other social science fields also provide guidance, but not generally descriptive and predictive behavior.

Econometrics provides models of large-scale economic activity as derived from past behavior. It is statistically based and usually operates by trying to discover models in the data rather than imposing models on data. In contrast, general system dynamics** builds dynamic models of social behavior by analysis of what linkages should be present, and then tests their aggregated models against history. System dynamics attempts to find large-scale behavioral patterns that are robust to the quantitative details of the model internals. Econometrics tries to make better quantitative predictions without having an avenue to abstract larger-scale structural behavior.

Experimental economics and equilibrium theory try to discover and manipulate a population's behavior in markets through use of microeconomic theory. As a real example, groups have applied these methods to pricing strategies for pollution licenses. Instead of setting pollution regulations economists have argued that licenses to pollute should be auctioned. This would provide control over the allowed pollution level (by the number of licenses issued) and be economically efficient. This strategy has been implemented in some markets and the strategies for conducting the auctions were tested by experimental groups beforehand. The object is to produce an auction system which results in stable equilibrium price for the licenses.

Law and economics is a branch of legal studies that applies micro- and macroeconomic principles to the analysis of legal and policy issues. It endeavors to assure economic efficiency in policies and to find least cost strategies to fulfill political goals. Although the concepts have gained fairly wide acceptance, they are inherently limited to those

---

* Rechtin, E., *Systems Architecting, Creating & Building Complex Systems*, Prentice-Hall, Englewood Cliffs, NJ, 1991, 188.
** An introductory reference on system dynamics is Wolstenholme, E. F., *System Enquiry: A System Dynamics Approach*, Wiley, Chichester, 1990, which explains the rationale, gives examples of application, and references the more detailed writings.

policy areas for market distribution which are considered politically acceptable.

Conclusions

A variety of powerful integrated modeling methods already exist in large domains. These methods exhibit, more or less explicitly, the progressions of refinement and evaluation noted as the organizing principle of architecting. In some domains, such as software, the models are very well organized, cover a wide range of development projects, and include a full set of views. However, even in these domains the models are not in very wide use and have less than complete support from computer tools. In some domains, such as sociotechnical systems, the models are much more abstract and uncertain. But in these domains the abstraction of the models matches the relative abstraction of the problems (purposes) and the systems built to fulfill the purposes.

Exercises

1. For a system familiar to you, investigate the models commonly used to architecturally define such systems. Do these models cover all important views? How are the models integrated? Is it possible to trace the interaction of issues from one model to another?

2. Build an integrated model of a system familiar to you covering at least three views. If the models in any view seem unsatisfactory, or integration is lacking, investigate other models for those views to see if they could be usefully applied.

3. Choose an implementation technology extensively used in a system familiar to you (software, board level digital electronics, microwaves, or any other). What models are used to specify a system to be built. That is, what are the equivalents of buildable blueprints in this technology. What issues would be involved in scaling those models up one level of abstraction so they could be used to specify the system before implementation design?

4. What models are used to specify systems (again, familiar to you) to implementation designers? What transformations must be made on those models to specify an implementation? How can the two levels be made better integrated?

Notes and References

1. White, S. et. al., Systems engineering of computer-based systems, *IEEE Computer*, p. 54, November 1993.

2. Maier, M. W., Quantitative engineering analysis with QFD, *Quality Eng.,* 7(4), 733, 1995.

3. Hauser, J. R. and Clausing, D., The house of quality, *Harvard Bus. Rev.,* 66(3), 63, May–June 1988.

4. The above-mentioned 1995 as well as Maier, M. W., Integrated modeling: a unified approach to system engineering, *J. Syst. Software,* February 1996.

5. Yourdon and Constantine, *Structured Design: Fundamentals of a Discipline of Computer Program and Systems Design,* Yourdon Press, Englewood Cliffs, NJ, 1979

6. DeMarco, T., *Structured Analysis and System Specification,* Yourdon Press, Englewood Cliffs, NJ, 1979.

7. DeMarco, T., *Controlling Software Projects,* Yourdon Press, Englewood Cliffs, NJ, 1982.

8. ADARTS Guidebook, SPC-94040-CMC, Version 2.00.13, Vol. 1–2, September 1991. Available through the Software Productivity Consortium, Herndon, VA.

9. Ward, P. T. and Mellor, S. J., *Structured Development for Real-Time Systems,* Vol. 1–3, Yourdon Press, Englewood Cliffs, NJ, 1985.

10. Rumbaugh, J. et. al., *Object-Oriented Modeling and Design,* Prentice-Hall, Englewood Cliffs, NJ, 1991.

11. Baudin, M., *Manufacturing Systems Analysis,* Yourdon Press Computing Series, Englewood Cliffs, NJ, 1990.

Part four

The systems architecting profession

The first three parts of this book have been about systems architecting as a process. This part is about systems architecting as a profession, that is, as a recognized vocation in a specialized field. Two factors are addressed here. The first, the political process,* is important because it interacts strongly with the architecting process, directly affecting the missions and designs of large-scale complex systems. The second, the professionalization of systems architecting, is important because it affects how systems architects as a group are perceived by the government, industry, and academia.

Chapter 10 is based on a course originated and taught at the University of Southern California by Dr. Brenda Forman of USC and the Lockheed Martin Corporation. The chapter describes the political process of the American government and the heuristics which characterize it. The federal process, instead of company politics or executive decision making,** was chosen for the course and for this book on architecting for three reasons.

First, federal governments are major sponsors *and clients* of complex systems and their development. Second, the American federal political process is a well-documented, readily available, open source for case studies to support the heuristics given here. And third, the process is assumed by far too many technologists to be uniformed, unprofessional, and self-serving. Nothing could be worse, less true, nor more damaging to a publicly supported system and its creators than acting under such assumptions. In actuality, the political process is the properly constituted and legal mechanism by which the general public expresses its judgments on the value to it of the goods and services that it needs. The fact that the process is time consuming, messy, litigious, not always fair to all, and certainly not always logical in a technical sense is far more a consequence of inherent conflicts of interests and values of the general public than of base motives or intrigue of its representatives in government.

The point that has been made many times in this book is that value judgments must be made by the client — the individual or authority that pays the bills — and not by the architect. For public services in

---

* By "political process" is meant the art and science of government, especially the process by which it acquires large-scale, complex systems.
** Company politics were felt to be too company specific, too little documented, and too arguable for credible heuristics. Executive decisions are the province of decision theory and are best considered in that context.

representative democratic countries, that client is represented by the legislative, and occasionally the judicial branch of the government.* Chapter 10 states a number of heuristics, the "facts of life", if you will, describing how that client operates. In the political domain, those rules are as strong as any in the engineering world. The architect should treat them with at least as much respect as any engineering principle or heuristic. For example, one of the facts of life states:

> **The best engineering solutions are not necessarily the best political solutions.**

Ignoring such a fact is as great a risk as ignoring a principle of mathematics or physics — one can make the wrong moves and get the wrong answers.

Chapter 11 addresses the challenge in the Preface to this book to *professionalize* the field; that is, to establish it as a profession** recognized by its peers and its clients. In university terms, this means at least a graduate-level, specialized education, successful graduates, peer-reviewed publications, and university-level research. In industry terms, it means the existence of acknowledged experts and specialized techniques. Dr. Elliott Axelband*** reports on progress toward such professionalization by tracing the evolution of systems standards toward architectural guidelines, by describing architecture-related programs in the universities, and by indicating professional societies and publications in the field. Axelband concludes the book with an assessment of the profession and its likely future.

---

* In the U.S., the executive branch *implements* the value judgments made by the Congress unless the Congress expressly delegates certain ones to the executive branch.

** "Any occupation or vocation requiring training in the liberal arts or the sciences and advanced study in a specialized field." *Webster's II New Riverside University Dictionary,* Riverside Publishing, Boston, MA, 1984, p. 939.

*** Associate Dean, School of Engineering, University of Southern California, and the director of the Systems Architecting and Engineering program. Dr. Axelband previously was an executive of the Hughes Aircraft Company until his retirement in early 1994.

chapter ten

The political process and systems architecting*

Brenda Forman

Introduction: the political challenge

The process of systems architecting requires two things above all others — value judgments by the client and technical choices by the architect. The political process is the way that the general public, when it is the end client, expresses its value judgments. High-tech, high-budget, high-visibility, publicly supported programs are therefore far more than engineering challenges; they are *political* challenges of the first magnitude. A program may have the technological potential of producing the most revolutionary weapon system since gunpowder, elegantly engineered and technologically superb, but if it is to have any real life-expectancy or even birth, its managers must take its political element as seriously as any other element. It is not only possible but likely that the political process will not only drive such design factors as safety, security, produceability, quantity, and reliability, but may even influence the choice of technologies to be employed. The bottom line is

If the politics don't fly, the system never will.

---

* This chapter is based on a course originated and taught at the University of Southern California by Dr. Forman, now with the Lockheed Martin Corporation in Washington, D.C. As indicated in the Introduction to Part 4, the political process of the American Federal Government was chosen for the course, for three reasons: it is the process by which the general public expresses its value judgments as a customer, it is well documented and publicized, and it is seriously misunderstood by the engineering community, to the detriment of effective architecting.

Politics as a design factor

Politics is a determining design factor in today's high-tech engineering. Its rules and variables must be understood as clearly as stress analysis, electronics, or support requirements. However, its rules differ profoundly from those of Aristotelian logic. Its many variables can be bewildering in their complexity and often downright ornery.

In addition to the formal political institutions of the Congress and the White House, a program must deal with a political process that includes interagency rivalries, intra-agency tensions, dozens of lobbying groups, influential external technical review groups, powerful individuals both within and outside government, and always and everywhere, the media.

These groups, organizations, institutions, and individuals interact in a process of great complexity. This confusing and at times chaotic activity, however, determines the budgetary funding levels that enable the engineering design process to go forward. More often of late, it directly affects the design in the form of detailed budget allocations, assignments of work, environmental impact statements, and reporting risk threats of outright cancellation.

Understanding the political process and dealing successfully with it is therefore crucial to program success.

> Example: Perhaps no major program has seen as many cuts, stretchouts, reviews, mandated designs, and risk of cancellation as the planetary exploration program of the 1970s and 1980s. Much of the cause was the need to fund the much larger Shuttle program. For more than a decade there were no planetary launches and virtually no new starts. From year to year changes were mandated in spacecraft design, the launch vehicles to be used, and even the planets and asteroids to be explored. The collateral damage to the planetary program of the Shuttle Challenger loss was enormous in delayed opportunities, redesigns, and wasted energy. Yet the program was so engineered that it still produced a long series of dramatic successes, widely publicized and applauded, using spacecraft designed and launched many years before.

Begin by understanding that *power is very widely distributed in Washington.* There is no single, clear-cut locus of authority to which to turn for support for long-term, expensive programs. Instead support must be continuously and repeatedly generated from widely varying groups, each of which may perceive the program's expected benefits in quite different ways and many of whose interests may diverge rather sharply when the pressure is on.

Example: The nation's space program is confronted with extraordinary tensions, none of which are resolvable by any single authority, agency, branch, or individual. There are tensions between civilian and military, between science and application, between manned and unmanned flight, between complete openness and the tightest security, between the military services, between NASA centers, and between the commercial and government sectors, to name a few. Typical of the contested issues are launch vehicle development, acquisition, and use; allocations of work to different sections of the country and the rest of the world; and of the future direction of every program. No one, anywhere, has sufficient authority to resolve any of these tensions and issues, much less to resolve them all simultaneously.

This broad dispersion of power repeatedly confuses anyone expecting that somebody will really be in charge there. Rather the opposite is true: anything that happens in Washington is the result of dozens of political vectors, all pulling in different directions. Everything is the product of maneuver and compromise. When those fail, the result is policy paralysis — and all too possibly, program cancellation by default or failure to act.

There are no clear-cut chains of command in the government. It is nothing like the military or even like a corporation. The process gets even more complicated because *power does not stay put* in Washington. Power relationships are constantly changing, sometimes quietly and gradually, at other times suddenly, under the impact of a major election or a domestic or international crisis. These shifts can alter the policy agenda — and therefore funding priorities — abruptly and with little advance warning. A prime example is the ever-changing contest over future defense spending levels in the wake of the welcomed end of the cold war.

The entire process is far better understood in dynamic than static terms. There is a continuous ebb and flow of power and influence between the Congress and the White House, among and within the rival agencies, and among ambitious individuals. And through it all, everyone is playing to the media, particularly to television, in efforts to change public perceptions, value judgments, and support.

The first skill to master

To deal effectively with this process, *the first skill to master is the ability to think in political terms.* And that requires understanding that the political process functions in terms of an entirely different logic system than the one in which scientists, engineers, and military officers are trained. Washington functions in terms of the logic of politics. It is a system every bit as rigorous in its way as any other, but its premises and rules are profoundly different. *It will therefore repeatedly arrive at conclusions quite different from those of engineering logic, based on the same data.*

Scientists and engineers are trained to marshall their facts and proceed from them to proof: for them, proof is a matter of firm assumptions, accurate data, and logical deduction. Political thinking is structured entirely differently. It depends not on logical proof, but on past experiences, negotiation, compromise, and perceptions. Proof is a matter of "having the votes." If a majority of votes can be mustered in Congress to pass a program budget, then — by definition — the program has been judged to be worthy, useful, and beneficial to the nation. If the program cannot, then no matter what its technological merits, the program will lose out to other programs which can.

Mustering the votes depends only in part on engineering or technological merit. These are always important — but getting the votes frequently depends as much or even more on a quite different value judgment, the program's benefits in terms of jobs and revenues among the congressional districts.

> Example: After the Lockheed Corporation won NASA's Advanced Solid Rocket Motor (ASRM) program, the program found strong support in the Congress because Lockheed located its plant in the Missisippi district of the then Chairman of the House Appropriations Committee. Lockheed's factory was only partially built when the chairman suffered a crippling stroke and was forced to retire from his congressional duties. Shortly thereafter, the Congress, no longer obliged to the chairman, reevaluated and then cancelled the program.

In addition to the highest engineering skills, therefore, the successful architect–engineer must have at least a basic understanding of this political process. The alternative is to be repeatedly blindsided by political events — and worse yet, not even to comprehend why.

Heuristics in the political process: "the facts of life"

Following are some basic concepts for navigating these rocky rapids — "The Facts of Life". They are often unpleasant for the dedicated engineer but they are perilous to ignore. Understanding them, on the other hand, will go far to help anticipate problems and cope more effectively with them. They are as follows and will be discussed in turn:

- Politics, not technology, sets the limits of what technology is allowed to achieve.
- Cost rules.
- A strong, coherent constituency is essential.

- Technical problems become political problems.
- The best engineering solutions are not necessarily the best political solutions.

Fact of life #1

Politics, not technology, sets the limits of what technology is allowed to achieve.

If funding is unavailable for it, any program will die, and getting the funding — not to mention keeping it over time — is a political undertaking. Furthermore, funding — or rather the lack of it — sets limits that are considerably narrower than what our technological and engineering capabilities could accomplish in a world without budgetary constraints. *Our technological reach increasingly exceeds our budgetary grasp.* This can be intensely frustrating to the creative engineer working on a good and promising program.

> Example: The space station program can trace its origins to the mid-1950s. By the early 1960s it was a preferred way station for traveling to and from the moon. But when, for reasons of launch vehicle size and schedule, the Apollo program chose a flight profile that bypassed any space station and elected instead a direct flight to lunar orbit, the space station concept went into limbo until the Apollo had successfully accomplished its mission. The question then was, what next in manned spaceflight? A favored concept was a manned space station as a way point to the moon and planets, built and supported by a shuttle vehicle to and from orbit. Technologically, the concept was feasible; some argued that it was easier than the lunar mission. Congress balked. The President was otherwise occupied. Finally, in 1972, the Shuttle was born as an overpromised, underbudgeted fleet, without a space station to serve. Architecturally speaking, major commitments and decisions were made before feasibility and desirability had been brought together in a consistent whole.

Fact of life #2

Cost rules.

High technology gets more expensive by the year. As a result, the only pockets deep enough to afford it are increasingly the government's. The fundamental equation to remember is **Money = Politics**. While reviews and hearings will spend much time on presumably technical issues, the fundamental and absolutely determining consideration is always affordability — and **affordability is decided by whichever side has the most votes.**

Funding won in one year, moreover, does not stay won. Instead it must be fought for afresh every year. With exceedingly few exceptions, no program in the entire federal budget is funded for more than 1 year at a time. Each and every year is therefore a new struggle to head off attackers who want the program's money spent somewhere else, to rally constituents, to persuade the waverers, and, if possible, to add new supporters.

This is an intense, continuous, and demanding process requiring huge amounts of time and energy. And after 1 year's budget is finally passed, the process starts all over again. There is always next year. Keeping a program "sold", in short, is a continuous political exercise, and like the heroine in the old movie serial, "The Perils of Pauline", some programs at the ragged edge will have to be rescued from sudden death on a regular basis. Rescue, if and when, may be only partial — not every feature can or will be sustained. If one of the lost features is a *system* function, the end may be near.

Example: After the Shuttle had become operational, the question again was, what next in manned spaceflight? Although a modestly capable space station had been successfully launched by a Saturn launch vehicle, the space station program had otherwise been shelved once the Shuttle began its resource-consuming development. With developmental skill again available, the space station concept was again brought forward. However, order-of-magnitude life cycle cost estimates of the proposed program placed the cost at approximately that of the Apollo, which in 1990-decade dollars would have been about $100 billion — clearly too much for the size of constituency it could command. The result has been an almost interminable series of designs and redesigns, all unaffordable as judged by congressional votes. Even more serious, the cost requirement has resulted in a spiraling loss of system functions, users, and supporters. Microgravity experiments, drug testing, onboard repair of on-orbit satellites, zero-g manufacturing, optical telescopes, animal experiments, military research and development — one after another had to be reduced to the point of lack of interest by potential users. A clearly implied initial purpose of the space station, to build one because the Soviet Union had one, was finally put to rest with the U.S. government's decision to bring Russia into a joint program with the U.S., Japan, Canada, and the European Space Agency. Another space station redesign (and probably a new launch program design) is likely; however, the uncertainty level in the program remains high, complicating that redesign. One apparent certainty: the U.S. Congress has made the value judgment that a yearly cap of U.S. $2.1 billion is all that a space station program is worth. The design must comply or risk cancellation. Cost rules.

Fact of life #3

A strong, coherent constituency is essential.

No program ever gets funded solely — or even primarily — on the basis of its technological merit or its engineering elegance. By and large, the Congress is not concerned with its technological or engineering content (unless, of course, those run into problems — see Fact of Life #4). Instead, program funding depends directly on the strength and staying power of its supporters, i.e., its *constituency*.

Constituents support programs for any number of reasons, from the concrete to the idealistic. At times, the reasons given by different supporters will even seem contradictory. From the direct experience of one of the authors, some advocates may support defense research programs because they are building capability; others because research, in promising better systems in the future, permits reduction if not cancellation of present production programs.

> Example: The astonishing success of the V-22 tilt-rotor Osprey aircraft program in surviving 4 years of hostility during the 1988–1992 period is directly attributable to the strength of its constituency, one that embraced not merely its original Marine Corps constituency but other armed services as well — plus groups that see it as benefiting the environment (by diminishing airport congestion), as improving the balance of trade (by tapping a large export market), and as maintaining U.S. technological leadership in the aerospace arena.

Assembling the right constituency can be a delicate challenge because a constituency broad enough to win the necessary votes in Congress can also easily fall prey to internal divisions and conflicts. Such was the case for Shuttle and is for the Space Station. The scientific community proved to be a poor constituency for major programs; the more fields that were brought in, the less the ability to agree on mission priorities. On the other hand, a tight homogeneous constituency is probably too small to win the necessary votes. The superconducting supercollider proved to be such. The art of politics is to knit these diverse motivations together firmly enough to survive successive budget battles and keep the selected program funded. Generally speaking, satellites for national security purposes have succeeded in this political art. It can require the patience of a saint coupled with the wiliness of a Metternich, but such are the survival skills of politics.

Fact of life #4

Technical problems become political problems.

In a high-budget, high-technology, high-visibility program, **there is no such thing as a purely technical problem**. Program opponents will be constantly on the lookout for ammunition with which to attack the program and technical problems are tailor-made to that end. And the problems will normally be reported in a timely fashion. As many programs have learned, **mistakes are understandable, failing to report them is inexcusable**. In any case, reviews are mandated by the Congress as natural part of the program's funding legislation. Any program that is stretching the technological envelope will inevitably encounter technical difficulties at one stage or another. The political result is that "technical" reports inevitably become political documents as opponents berate and advocates defend the program for its real or perceived shortcomings.

Judicious damage prevention and control, therefore, are constantly required. Reports from prestigious scientific groups such as the NRC or DSB will routinely precipitate congressional hearings in which hostile and friendly congressmen will pit their respective expert witnesses against one another, and the program's fate may then depend not only on the expertise, but on the political agility and articulateness of the supporting winesses. Furthermore, while such hearings will spend much time on ostensibly technical issues, the fundamental and absolutely determining consideration is always affordability — and affordability is decided by whichever side has the most votes.

> Examples: Decades-long developments are particularly prone to have their technical problems become political. Large investments have to be made each and every year before any useful systems appear. The widely reported technical difficulties of the B-1 and B-2 bombers, the C-17 cargo carrier, the Hubble telescope, and the Galileo Jupiter spacecraft became matters of public as well as legislative concern. The future of long distance air cargo transport, of space exploration, and even of the NASA are all brought up for debate and reconsideration every year. Architects, engineers, and program managers have good reason to be concerned.

Fact of life #5

The best engineering solutions are not necessarily the best political solutions.

Remember that we are dealing with two radically different logic systems here. *The requirements of political logic repeatedly run counter of those of engineering logic.* Take construction schedules: in engineering terms, an optimum construction schedule is one that makes the best and most economical use of resources and time and yields the lowest unit cost. In political terms, the optimum construction schedule is the one that the political process decides is affordable in the current fiscal year. These two definitions routinely collide; the political definition always wins.

> Example: NASA and other agencies often refer to what is called the program cost curve. It plots total cost of development and manufacture as a function of its duration, as follows:

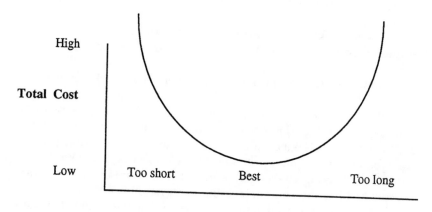

Duration in Years

The curve is logical and almost always true. But it is irrelevant because *the government functions on a cash-flow basis.* Long-term savings will almost always be foregone in favor of minimizing immediate outlays. Overall life-cycle economies of scale will repeatedly be sacrificed in favor of slower acquisitions and program stretchouts because these require lower yearly appropriations, even if they cause higher unit costs and greater overall program expense. There is also the contradictory perception that if a given program is held to a tight, short schedule it will cost less — facts not withstanding. (See Chapter 5, Social Systems, Facts vs. Perceptions.)

The foregoing example leads to another provisional heuristic:

With few exceptions, schedule delays are accepted, albeit grudgingly; cost overruns are not, and for good reason.

The reason is basic. A cost overrun, that is, an increase over budget in a given year, will force the client to take the excess from some other program and that is not only difficult to do, it is hard to explain to the blameless loser and to that program's supporters. Schedule delays mean postponing benefits at some future cost — neither of which affect anyone today.

> Example: Shuttle cost overruns cost the unmanned space program and its scientific constituency two decades of unpostponable opportunities, timely mission analyses, and individual careers based on presidentially-supported, wide-consensus planning.

By the same token, a well-run program that sticks to budget can encounter very difficult technical problems and survive.

> Examples: Communication and surveillance satellite programs.

All in all, it can be a bewildering and intimidating process to the uninitiated. But it need not be so, because, in addition to being confusing and chaotic, this is a profoundly interesting and engrossing process, every bit as challenging as the knottiest engineering problem. Indeed it *is* an engineering challenge because it molds the context in which systems architecting and engineering must function.

The reader may well find the craziness of the political process distasteful — *but it will not go away.* The politically naive architect may experience more than a fair share of disillusion, bitterness, and failure. The politically astute program manager, on the other hand, will understand the political process, will have a strategy to accommodate it, and will counsel the architect accordingly. Some suggestions for the architect: it helps to document accurately when and why less-than-ideal technical decisions were made — and how to mitigate them later, if necessary. It helps to budget for contingencies and reasonably foreseeable risks. It helps to have stable and operationally useful interim configurations and fallback positions. It helps to acknowledge the client's right to have a change of mind or to make difficult choices without complaint from the supplier. Above all, it helps to acknowledge that living in the client's world can be painful, too. And finally, select a kit of prescriptions for the pain such as the following from Appendix A:

- The triage, when there is only so much that can be done: let the dying die. Ignore those who will recover on their own. And treat only those who would die without help.
- The most important single element of success is to listen closely to what the customer (in this case, the Congress) perceives as his requirements and to have the will and ability to be responsive (J.E. Steiner, The Boeing Company, 1978).
- Look out for hidden agendas.
- And so on.

That doesn't mean that architects and engineers have to become expert lobbyists — but it does mean having an understanding of the political context within which programs must function, the budget battle's rules of engagement, and of those factors that are conducive to success or failure. The political process is not outside, it is *an essential element of* the process of creating and building systems.

A few more skills to master

Following are a few more basic coping skills for the successful systems architect. First and foremost, *understand that the Congress and the political process are the <u>owners</u> of your project. They are the ultimate <u>clients</u>.* It is absolutely essential to deal with them accordingly by making sure they understand what you are trying to do, why it is important, and why it makes political sense for them to support you.

Be informed. This is your life, so be active. Learn the political process for yourself and keep track of what's going on. Figure out what information the political system needs in order to understand what the program needs — and arrange to supply it to them. A chief engineer has utterly different information requirements from a congressional oversight committee. Learn what sort of information furthers your program's fortunes in Washington and then get it to your program managers so they can get it to the political decision makers who determine your program's funding. Maybe your program has a great job-multiplier effect in some crucial lawmaker's district. Maybe its technology has some great potential commercial applications in areas where the U.S. is losing a competitive battle with another country.

The point is that *the political process bases its decisions on very different information than the engineering process does.* Learn to satisfy both those sets of requirements *by plan.*

Conclusion

The political process is a necessary element of the process of creating and building systems. It is not incomprehensible; it is different. Only when they are not understood do the political facts of life instill cynicism or a sense of powerlessness. Once understood, they become tools like any others in the hands of an astute architect. It is a compliment to the client to use them well.

chapter eleven

The professionalization of systems architecting*

Elliott Axelband

> *Profession: Any occupation or vocation requiring training in the liberal arts or the sciences and advanced study in a specialized field.*[1]

Introduction

To readers who have progressed this far, the existence of systems architecting as a *process*, regardless of who performs it, can be taken for granted. Functions and forms have to be matched, system integrity has to be maintained throughout development, and systems have to be certified for use. Visions have to be created, realized, and demonstrated.

This chapter, in contrast, covers the evolution of the systems architecting *profession*. An appropriate place to begin is with the history of the closely related profession of systems engineering, the field from which systems architecting evolved.

The profession of systems engineering

Systems engineering as a process began in the early 1900s in the communication and aircraft industries. It evolved rapidly during and after World War II as an important contributor to the development of large, innovative, and increasingly complex systems. By the early 1950s systems engineering had reached the status of a recognized, valued profession. Communication networks, ballistic missiles, radars, computers, and satellites were all recognized as systems. The "systems

---

* This chapter is based on research by Dr. Axelband, Associate Dean of Engineering of USC, and formerly an executive at Hughes Aircraft Company. It recognizes that if systems architecting is to be mature and effective, it must be or become a true profession.

approach" entered into everyday language in many fields, social as well as technical. Government regulations and standards explicitly addressed systems issues and techniques. Thousands of engineers called systems engineering their vocation. Professional societies formed sections with journals devoted to systems and their development.[2] Universities established systems engineering departments or systems-oriented programs.* Books addressing the process, or aspects of it, started to appear.[3] Most recently, the profession became formally represented with the establishment of the International Council on Systems Engineering (INCOSE).[4]

The core of the system approach from its beginnings has been the definition and management of system interfaces and trade-offs, of which there can be hundreds in any one system. System analysis, system integration, system test, and computer-aided system design were progressively developed as powerful and successful problem-solving techniques. Some have become self-standing professions of their own under the rubric of systems engineering. Their academic, industrial, and governmental credentials are now well established.

All are science based, that is, on measurables and a set of assumptions. In brief, these are that requirements and risks can be quantified, solutions can be optimized, and compliance specified. But these same assumptions are also constraints on the kinds of problems that can be solved. In particular, science-based systems engineering does not do well in problems that are abstract, data deficient, perceptual, or for which the criteria are unmeasurable.

For example, the meanings of such words as safe, survivable, affordable, reliable, acceptable, and good and bad are either outside the scope of systems engineering — "ask the client for some numbers" — or are force-fitted to it by subjective estimates. Yet these words are the language of the clients. Quantifying them can distort their inherent vagueness into an unintended constraint.

There is no group of professionals which better understands these difficulties than systems engineers and executives — nor who wish more to convert unmeasurable factors to quantitatively statable problems by whatever techniques can help. The first step they made was to recognize the nature of the problems. The second was to realize that almost all of them occur at the front (and back) ends of the engineering cycle. Consider the following descriptive heuristics, developed long ago from systems engineering experience:

- All the serious mistakes are made in the first day.

---

* Among the best known are the University of Arizona at Tucson, Boston University, Carnegie Tech, Case Western Reserve, the University of Florida, Georgia Tech, the University of Maryland, George Mason University, Ohio State, MIT (Aerospace), New York Polytech, the University of Tel Aviv, the University of Southern California, Virginia Polytechnic Institute, and the University of Washington.

- Regardless of what has gone before, the acceptance criteria determine what is actually built.
- Reliability has to be designed in, not tested in.

It is no coincidence that many systems engineers, logically enough, now consider systems architecting to be "the front end of systems engineering" and that architectures are "established structures." More precisely, systems architecting can be seen as setting up the necessary conditions for systems engineering and certifying its results. In short, systems architecting provides concepts for analysis and criteria for success.

The immediate incentive for making architecting an explicit process, the necessary precursor to establishing it as a self-standing profession complementary to systems engineering, was the recognition in the late 1980s by system executives that "something was missing" in system development and acquisition; and that the omission was causing serious trouble: system rejection by users, loss of competitive bids to innovators, products stuck in unprofitable niches, military failures in joint operations, system overruns in cost and schedule, and so on — all traceable to their beginnings. Yet there was a strong clue to the answer. Retrospective examinations of earlier, highly successful systems showed the existence in each case of an overarching vision, created and maintained by a very small group, that characterized the program from start to finish.[5]

Software engineers and their clients were among the first to recognize that the source of many of their software system problems was structural, that is, architectural.* Research in software architecture followed in such universities as Carnegie Mellon, the University of North Carolina Chapel Hill, the University of California at Irvine, and the University of Southern California. Practitioners began identifying themselves as software architects and forming architectural teams. Communication, electronics, and aerospace systems architects followed shortly thereafter.

Societies established architecture working groups, notably the INCOSE Systems Architecting Working Group[6] and the IEEE Software Engineering Standards Committee's Architecture Working Group,[7] to formulate standard definitions of terms and descriptions for systems and software architectures. These activities are essential to the development both of a common internal language for systems architecting

---

* One of the earliest and most famous books on systems architecting is *The Mythical Man-Month, Essays on Software Engineering* by Frederick P. Brooks, Jr. (Addison-Wesley Publishing Company, 1974), which not only recognized the structural problems in software, but explictly, on p. 37, calls for a system architect, a select team, conceptual integrity, and for the architect to "confine himself scrupulously to architecture" and to stay clear of implementation. Brook's analogy for the architectural team was a surgical team. He credits a 1971 Harlan Mills proposal as the source of these precepts.

and for the integration of software and systems architectures in complex, software-intensive systems.

At the scale of the profession of engineering, the recognition that something was missing led to identifying it, by direct analogy with the processes of the classical architectural profession, as systems architecting.[8] Not surprisingly, the evolution of systems architecting tools was found to be already underway in model building, discussed in Part 3, and system standards,* discussed in the next section.

Systems architecting and system standards

Earlier chapters have pointed out that the abstract problems of the conceptual and certification phases require different tools from the analytic ones of system development, production, and test. One of the most important sets of tools is that of system standards. For historical reasons, *architectural* system architectural standards were not developed as a separate set. Instead, general systems standards were developed which included systems architecting elements and principles understood at the time, most of them induced from lessons learned in individual programs. As will be seen, some key elements appeared as early as the 1950s

Driven by much the same needs, the recognition of systems architecting in the late 1980s was paralleled, independently, by a recognition that existing system standards needed to be modified or supplemented to respond to long-standing systems-level structuring problems. Bringing the two tracks, architecting and standards, together should soon help both. Architecting can improve system standards. System standards can provide valuable tools for the system architecting profession.

Some of the earliest system standards in which elements of systems architecting appeared were those of system specification, interface description, and interface management. They proliferated rapidly. A system specification can beget ten subsystem specifications, each of which is supported by ten lower-level subsystem specifications, and so on. All of these had to be knitted together by a system of 100 or so interface specifications.

Even though modern computer tools (computer-assisted system engineering or CASE tools) have been developed to help keep track of the systems engineering process, extraordinarily disciplined efforts are still required to maintain the system's integrity.[9]

As system complexity increased, systems engineers were faced with increasingly difficult tasks of assuring that the evolving form of the system met client needs, guaranteeing that trade-offs maintained

---

* Systems standards, for the purposes of this book, are those engineering standards having impact on the system as a whole, whether explictly identified as such in their titles or not. They are a relatively small part of the totality of engineering standards. Many, if not most, are interface and test standards.

system intent in the face of complications arising during development, and finally assuring that the system was properly tested and certified for use. In due course, the proliferation of detailed specifications led to a need for overarching guidelines, an overview mechanism for "structuring" the complexity that had begun to obscure system intent and integrity.

Before continuing it should be pointed out that overarching guidelines are not, and cannot be, a *replacement* for quantitative system standards and specifications. The latter represent decades of corporate memory, measurable acceptance criteria, and certified practices. Guidelines — performance specifications, tailorable limits, heuristics, and the like — have a fundamental limitation. They cannot be certified by measurables. They are too "soft" and too prone to subjective perceptions to determine to the nearest few percentage points whether a system performs or costs what it should. At some point, the system has to be measured if it is to be judged objectively.

From the standpoint of an architecting profession, the most important fact about system standards is that they are changing. To understand the trend, their development will be reviewed in some detail, recognizing that some of them are continuing to be updated and revised.

The origins of system standards

The ballistic missile program of the 1950s

Urgent needs induce change, and — eventually — improvement. The U.S./Soviet ballistic missile race begun in the mid- to late 1950s brought about significant change, as it led to the development and fielding of innovative and complex systems in an environment where national survival was threatened. To its credit, the U.S. Air Force recognized the urgent need to develop and manage the process of complex system evolution, and did so.* The response in the area of standards was the "375" system standard, subsequently applied to the development of all new complex Air Force equipment and systems.

"375" required several things which are now commonplace in systems architecting and engineering. Time lines depicting the time-sequenced flow of system operation were to be used as a first step in system analysis.** From these, system functional block diagrams and

---

* Those responsible for this development, Dr. Simon Ramo and General Bernard Schriever, in particular, from time to time referred to their respective organizations as architects as well a system integrators. "Architecture," as a formalism, was largely bypassed in the urgency to build balllistic missiles as credible deterrents. Nonetheless, the essential "architectural" step of certification of readiness for launch was incorporated from the beginning and executed by all successor organizations. It became a centerpiece for the space launch programs of the 1960s and thereafter.

** This is not necessarily appropriate for all systems, but it was well suited to the missile, airplane, and weapon systems the Air Force had in mind at the time.

functional requirements were to be derived as a basis for subsequent functional analysis and decomposition. The functional decomposition process in turn generated the subsystems which with their connections and constraints comprised the system, and allowed the generation of subsystem requirements via trade-off processes.

"375" was displaced in 1969 by a MILSTD 499 (Military Standard–499)* which was applied throughout the Department of Defense. MILSTD 499A, an upgrade, was released in 1974 and was in effect for 20 years. MILSTD 499B, a later upgrade, was unofficially released in 1994, and was almost immediately replaced by EIA/IS 632 Interim Systems Engineering Standard.[10]

The beginning of a new era of standards

The era of MILSTDs 499/499A/499B was an era in which military standards became increasingly detailed. It was not only these documents which governed system architecting and engineering, but they in turn referenced numerous other DoD (Department of Defense)MILSTDs which addressed aspects of system engineering, and which were imposed on the military system engineering process as a consequence. To cite a few: MILSTD 490, Specification Practices, 1972; MILSTD 481A, Configuration Control–Engineering Changes, Deviations and Waivers (Short Form), 1972; MILSTD 1519, Test Requirements Document, 1977; and MILSTD 1543, Reliability Program Requirements for Space and Missile Systems, 1977. See Eisner, 1994[11] for additional examples.

This mind set changed with the end of the cold war in the late 1980s. Cost became an increasingly important decisive factor in competitions for military programs, supplanting performance which had been the dominant factor in the prior era. Lowest cost, it was argued, could only be achieved if the restrictions of the military standards were muted. The detailed process ("how to") standards of the past, which specified how to conduct systems engineering and other program operations, needed to be replaced by standards that only provided guidelines** leaving the engineering specifics to the proposing companies which would select these so as to be able to offer a low-cost product.[12] Further supporting this reasoning was the reality that the most sophisticated components and systems in fields such as electronic computer chips and computers were now available at low cost from commercial sources, whereas in the past the state of the art was only available from MILSTD-qualified sources. It was in this environment that EIA/IS 632 was born.

---

* The official form is "Mil. Std.–499" but for ease of reading in a text an alternate form, "MILSTD 499", will be used here.

** As noted earlier, *replacement* is a questionable motivation for guidelines. Nonetheless, the establishment of a high-level guideline document — a key architecting technique — was a milestone.

EIA/IS 632, an architectural perspective

EIA/IS 632 is short by comparison with other military standards. Its main body is 36 pages. Including appendices its total length is 60 pages, and these include several which have been left intentionally blank. And most significantly, no other standards are referenced.

The scope and intent of the document are best conveyed by the following quotes from its contents:

- "The scope .... of systems engineering (activities) is defined in terms of what should be done, not how to do ... (them).." p. i
- " ... (EIA/IS–632) identifies and defines the systems engineering tasks that are generally applicable throughout the system life cycle for any program ...." p. 7

From a systems architecting perspective it is clear that the scope of the life-cycle perspective includes the modern understanding of systems architecting.

One of the major activities of the systems architect, that of giving form to function, is addressed in p. 9–11. These pages summarize, in their own words and style, the client/architect relationship, the establishment of the defining system functions, the development of the system's architecture, and the process of allocating system functions to architectural elements via trade-offs. By implication, the trade-offs continue, with varying degrees of concentration, throughout the life cycle.

Curiously, test and validation are deferred to a later section entitled "4.0 Detailed Requirements". This is consistent with the historical organization of the preceding military standards, wherein Section 4 was dedicated to product assurance, a term which included system test. It is, however, a significant departure from the systems architecting point of view. A basic tenet of systems architecting is that certification for use is one of its most important functions, and that this should be developed in parallel with — and as a part of — the development of a system's architecture. Consider, for example, some of the architecting heuristics that could apply:

- To be tested a system must be designed to be tested.
- Regardless of what has gone before, the acceptance (and test) criteria determine what is actually built.

There are other sound and basic architecting principles which, suitably explained, could and should have been included as historically validated guiding principles in EIA/IS 632 which, in its own words, "provides guidance for the conduct of a systems engineering effort". Some applicable heuristics would include:

- Simplify. Simplify. Simplify.

- The greatest leverage in systems architecting is at the interfaces.
- Except for good and sufficient reasons, functional and physical structuring should match.
- In partitioning a system into subsystems, choose a configuration with minimal communications between subsystems.
- It is easier to match a system to the human one that supports it than the reverse.

Beyond these, the need for an unbiased agent — the system architect — to represent the client and technically guide the process is absent and a serious omission.

Commercial standards

While EIA/IS 632 applies only to military systems engineering, that was not its original intent. The objective was to develop a universal standard for systems engineering which would apply to both the military and commercial worlds, and be ratified by all of industry. However, there was an urgency to publish a new military standard and in the 4-month schedule that was assigned, there was a limited amount of "remilitarization" that could be accomplished. This led to two consequences. First, IEEE P1220,[13] a commercial systems engineering standard, was separately published. Second, the merging of EIA/IS 632 with IEEE P1220 to create the first universal standard for system engineering was planned for publication in 1997. The development of this universal system engineering standard involves personnel from several organizations including EIA, IEEE, and INCOSE, and is expected to be named EIA632.[14] Beyond that, ISO (the International Standards Organization) plans to issue a standard in 2000+ which will merge EIA632 with ISO 12207. This latter will be published soon and will provide a universal software engineering standard. In passing, software is included in the definition of a system as used in the prior-mentioned systems engineering standards, but not with the same level of intensity as can be found in separate existing software engineering standards.

IEEE P 1220, an architectural perspective
The IEEE working group which generated P 1220 was sponsored by the IEEE Computer Society and included representatives from INCOSE, the EIA, and the IEEE AES Society. It is the first commercial standard to formally address systems engineering. P 1220's similarity with EIA/IS 632 derives from a fair degree of common authorship plus a deliberate effort to: (1) coordinate efforts in order to present a common view of systems engineering, and (2) anticipate the eventual merger of the two documents. While the similarities are therefore not surprising, there are significant differences which are worthy of mention.

To begin with similarities, both standards are guides and not "how to" instruction manuals. Both address the entire life cycle of a product. Both share a common architecture, addressing in similar ways things which are becoming similar — the processes of system engineering in the military and commercial environments. This extends to a fair degree of common vocabulary, although mercifully P 1220 is more free of acronyms.

Compared to EIA/IS 632, P 1220 is more complex and longer (58 pages in the body of the report vs. 36, and 66 pages overall vs. 60). It is much more rigorous in its definitions and use of system hierarchical structures. It has several significant differences which tend to favor the recognition and processes of systems architecting.

- The subsystems which comprise a system are understood and treated as systems. p. 2, 4, 13
- The customer (client) is explicitly identified along with a need to determine and quantify his expectations. p. 35
- External constraints including public and political constraints are recognized as part of the process. p. 35
- The role of system boundaries and constraints in system evolution is considered. p. 36
- The need to evolve test plans with product evolution is expressed. p. 18, 19
- The explicit need to generate functional and physical architectures is recognized, unfortunately (from a system architect's view) in the same section of the document which through usage defines system architecture as the sum of the product and its defining data package. p. A-3

In summary, P 1220 better recognizes the systems architecting process than does EIA/IS 632. It does however have significant systems architecting shortfalls and would better serve as a systems engineering guide if the role of the systems architect were included and if the architecting heuristics given in this chapter were added.

Company standards

Each company has its own set of standards and practices which incorporate unique core competencies, practices, and policies. These need to evolve for a company to improve its performance and competitive posture. Company standards serve two other functions: instructing its initiates and relating to its customers. The latter function is stimulated whenever customers change their standards, and it is from this perspective that the systems engineering standards of several companies were examined. This was not an easy task since systems architecting and engineering are viewed by those companies engaged in them as an enabler of efficient product generation, and as such applicable practices providing a competitive advantage are considered trade secrets.

Several generalizations are possible. Today's competitive pressures have caused self-examination and particularly re-engineering to become a regular way of life. This has also been encouraged by popularized business literature.[15] Process, as opposed to product, is the focus of such institutionalized activity. In reviewing the process of product generation and support, systems architecting and, in some cases, systems architects are gaining recognition, although not always in a way clearly separated from systems engineering.

The Harris Corporation Information Systems Division recently culminated 4 years of activity by publishing their revised *Systems Engineering Guide Book*. A generalized description is provided in Honour, 1993.[16] The 128-page book is company proprietary. Discussions with its author, Eric Honour, indicated that while systems architecting is not delineated per se, the *processes* that constitute systems architecting account for approximately 25% of its pages.

Sarah Sheard and Elliot Margolis recently reported on the evolving systems engineering process within the Loral Federal Systems Organization.[17] Their conclusion is that there is an important relationship between the nature of a product and the team developing it, and that as such there is no one best organization for product development. However, their recent experience indicates success with a team structure that includes a distinct architecture team with a clearly identified chief architect working in conjunction with a management team and both software and hardware development teams. Their use of the terms architect, architectures, and architecting are consistent with those of this book.

Hughes Aircraft published its 3 in.-thick *Systems Engineering Handbook* in 1994.[18] Its objective, stated on p. P-1, is to " .... improve both the quality and efficiency of systems engineering at Hughes". The handbook describes the then applicable MILSTD 499B and the Hughes systems engineering processes for both DoD and non-DoD programs, including Hughes' organizational and other resources available to implement these. It is a very comprehensive user-friendly book, clearly adapted from MILSTD 499B, and includes the activities of the systems architect — who is never identified by that name — within the framework of systems engineering. The systems engineering function and organization are identified as the technical lead organization for product development and provided a unique identity in all forms of organization discussed: functional, projectized, and matrix. In that the handbook is patterned after MILSTD 499B which has a strong resemblance to EIA/IS 632, the comments made earlier with respect to EIA/IS 632 apply.

A summary of standards developments, 1950–1995

For a variety of reasons and by a number of routes, system standards and specifications are evolving consistent with the principles and techniques of systems architecting. The next step is the use of system architects to help improve system standards, particularly in system conception, test, and certification. At the same time, improved system standards can provide powerful tools for the systems architecting profession.

A cautionary note: the recent and understandable enthusiasm in the Department of Defense to streamline standards and eliminate all references to prior MILSTDs could make systems architecting considerably more difficult. Useful as guidelines are, they are no substitute for quantitative standards for bidding purposes, for certifying a system for use, or for establishing responsibility and liability.* MILSTDs in many instances incorporate specific philosophical and quantitative requirements based on lessons dearly learned in the real world. They reduce uncertainty in areas that should not or need not be uncertain. To ignore these by omission is to run the risk of learning them all over again, at great cost. To the extent that the lessons relearned are architectural, the risks can be enormous. As the heuristic states, all the serious mistakes are made on the first day.

Systems architecting graduate education

Systems engineering universities and systems architecting

Graduate education, advanced study, and research give a profession its character. They distinguish it from routine work by making it a vocation, a calling of the particularly qualified.

The first university to offer masters and doctorate degrees in systems engineering was the University of Arizona, beginning in 1961. The program began as a graduate program; an undergraduate program and the addition of industrial engineering to the department title came later. Still in existence, the graduate department has well over 1000 alumni.

The next to offer advanced degrees was the Virginia Institute of Technology in 1971, but not until after 1984 did additional universities join the systems engineering ranks. They included Boston University, George Mason University, the Massachusetts Institute of Technology, the University of Maryland, the University of Southern California (USC), the University of Tel Aviv, and the University of Washington. It is worth noting that all are located at major centers of industry or government, the principal clients and users of systems engineering.

---

* In this connection, the Department of Defense has explictly retained interface and certification standards as essential, not to be considered as candidates for elimination.

To the best knowledge of the authors of this book, the University of Southern California is the first to offer a graduate degree in systems architecting and engineering with the focus on systems architecting. However, of the universities offering graduate degrees in systems engineering, some half dozen now include systems achitecting within their curricula. Architecting is also becoming a strong interest in universities offering advanced degrees in computer science with specializations in software and computer architectures; notably, Carnegie Mellon University, the Universities of California at Berkeley and Irvine, and USC. At USC, the systems architecture and engineering degree began with an experimental course in 1989, and formally became a masters degree program in 1993 following its strong acceptance by students and industry.

In the last 10 years there has been growing recognition of the value of interdisciplinary programs, which of itself would favor systems architecting and engineering. These have been soul-searching years for industry, and the value of system architecting and engineering has become appreciated as a factor in achieving a competitive advantage. Also, the restructuring of industry has caused a rethinking of the university as a place to provide industry-specific education. These trends, augmented by the success of the system architecting and engineering education programs, have caused university architect–engineering programs to prosper.

The success of these programs can be measured in several ways. First, the direction is one of growth. Seven out of the eight existing masters programs were started in the last 10 years. And the Universities of Maryland, Tel Aviv, and Southern California are all considering expanding their programs to include a Ph.D. Second, systems architecting and systems architecting education are making a positive difference in industry, as supported by industry surveys. In point of fact, company-sponsored systems architecting enrollments have increased in a period of industrial contraction.

Curriculum design

It is not enough in establishing a profession to show that universities are interested in the subject. The practical question is what is actually taught; that is, the curriculum. Because USC is apparently the first university to offer an advanced degree specifically in systems architecture and engineering, its curriculum is described. It should be pointed out that this curriculum is at the graduate level. To date, no undergraduate degree is offered or planned.

The USC masters program admits students satisfying the School of Engineering's academic requirements and having a minimum of 3 years applicable industrial experience. Students propose a ten-course curriculum which is reviewed, modified if required, and accepted as part of their admission. The curriculum requires graduate level courses as follows:

- An anchor systems architecting course
- An advanced engineering economics course
- 1 of several specified engineering design courses
- 2 elective courses in technical management from a list of 11 which are offered
- 1 of 8 general technical area elective courses
- 4 courses from 1 of 11 identified technical specialization areas, each of which has 6 or more courses offered

The structure of this MS in Systems Architecture and Engineering curriculum has been designed based on both industrial and academic advice. Systems architecture is better taught in context. It is too much to generally expect a student to appreciate the subtleties of the subject without some experience. And the material is best understood through a familiar specialty area in which the student already practices. The 3-year minimum experience requirement and the requirement of four courses in a technical specialty area derive from this reasoning.

The need for an anchor course is self-evident. Systems architecture derives from inductive and heuristic reasoning, unlike the deductive reasoning used in most other engineering courses. To fully appreciate this difference, the anchor course is taken early, if not first, in the sequence. The course contains no exams as such but requires two professional quality reports so that the student can best experience the challenges of systems architecting and architecture by applying his knowledge in a dedicated and concentrated way.

Experience has shown that a design experience course, the advanced economics course, and courses in technical management are valuable to the system architect, and therefore they are curriculum requirements. The additional course in a general technical area allows the student to select a course that most rounds out the student's academic experience. Possibilities include a systems architecting seminar, a course on decision support systems, and a course on the political process in systems architecture design.

To date, within the USC program there have been 28 masters graduates. Over 250 have taken the anchor course, and 30 students are enrolled masters candidates. A library of over 150 research papers has been produced from the best of the student papers and is available as a research library for those in the program. Several students are proceeding toward a Ph.D. in Systems Engineering with a specialty in Systems Architecting.

Advanced study in systems architecting

A major component of advanced study in any profession is graduate level research and refereed publications at major universities. In systems architecting advanced study can be divided into two relatively distinct parts: that of its science, closely related to that of systems engineering, and of its art. The universities committed to systems engineering education were given earlier. Advanced study in its art, though often illustrated by engineering examples, has many facets, including research in:

- Complexity, by Flood and Carson[19] at City University London, England
- Problem solving, by Klir[20] at the State University of New York at Binghamton and by Rubinstein[21] at the University of California at Los Angeles
- Systems and their modeling, by Churchman[22] at the University of California at Berkeley (UCB) and Nadler[23] and Rechtin[24] at the University of Southern California (USC)
- The behavioral theory of architecting, by Lang[25] at the University of Pennsylvania, Rowe[26] at Harvard, and Losk, Carpenter, Cureton, Geis, and Carpenter[27] at USC
- The practice of architecture, by Alexander[28] and Kostof[29] at UCB
- Machine (artificial) intelligence and computer science, by Genesereth and Nilsson[30] at Stanford, Newell[31] and Simon[32] at Carnegie Mellon University, and Brooks[33] at the University of North Carolina at Chapel Hill
- Software architecting, by Garlan and Shaw[34] at Carnegie Mellon University and Barry Boehm at USC

All have contributed basic architectural ideas to the field. Many are standard references for an increasing number of professional articles by a growing number of authors. Most deal explicitly with systems, architectures, and architects, although the practical art of systems architecting was seldom the primary motivation for the work. That situation predictably will change rapidly as both industry and government face international competition in a new era.

Professional societies and publications

Existing journals and societies were the initial professional media for the new fields of systems architecting and engineering. Since much early work was done in aerospace and defense, it is understandable that the IEEE Society on Systems Man and Cybernetics, the IEEE Aerospace Electronics Society, and the American Institute of Aeronautics and Astronautics, and their journals, along with others, became the professional outlets for these fields. One excellent sample paper from this period (Boonton and Ramo, 1984[35]) explained the contributions that systems engineering had made to the U.S. ballistic missile program.

The situation changed in 1990 when the first National Council on System Engineering conference was held and attracted 100 engineers. INCOSE became the first professional society dedicated to systems engineering and soon established a Systems Architecting Working Group.

The society, with a current membership of 2000, publishes a quarterly newsletter and a journal. The journal first appeared in 1994 and plans are in progress for the first of its annual joint publications with the IEEE AES Society in 1996. This significant step will impose IEEE reviewing standards and provide publication of systems architecting papers to a combined professional readership in excess of 9000. The journal will be placed in the libraries of most U.S. universities and a large number of universities outside the U.S. which subscribe to the *IEEE AES Transactions*.

Conclusion: an assessment of the profession

The profession of systems architecting has come a long way — and its journey has just begun. Its present body of professionals in industry and academia, beginning most often in electronics, control, and software systems, soon broadened into systems engineering, formed the core of small design teams, and now consider themselves as architects. The profession has been nurtured within the framework of systems engineering, and no doubt will maintain a tight relationship with it. A masters-level university curriculum now exists. Applicable research is underway in universities. Applicable standards and tools are being developed at the national level. It has an acknowledged home within INCOSE as well as other professional societies which, together with their publications, provide a medium for professional expression and development.

It is interesting to speculate on where the profession might be going and how it might get there. The cornerstone thought is that the future of a profession of systems architecting will be largely determined by the perceptions of its utility by its clients. If a profession is useful, it will be sponsored by them and prosper. To date, all indicators are positive.

Judging by the events that have led to its status today, and by comparable developments in the history of classical architecture, systems architecting could well evolve as a separate business entity. The future could hold more systems architecting firms that bid for the business of acting as the technical representative or agent of clients with their builders. There are related precedents today in Federally

Funded Research and Development Centers (FFRDCs) and Systems Engineering and Test Assistance Contractors (SETACs), independent entities selected by the Department of Defense to represent it with defense contractors that build end products. Similar precedents exist in NASA and the Department of Energy.

The role of graduate education is likely to grow and spread. Today's products and processes are more netted and interrelated than those of 10 years ago, and tomorrow's will be even more so. System thinking is proving to be fundamental to commercial success, and systems architecting will increasingly become a crucial part of new product development. It is incumbent upon universities to capture the intellectual content of this phenomenon and embody it in their curricula. This will require a tight coupling with industry to be aware of important real world problems, a dedication to research to provide some of the solutions, and an education program which trains students in relevant architectural thinking.

Published peer-reviewed research has stood the test of time, providing the best medium for the rapid dissemination of state-of-the-art thinking. Today INCOSE's Systems Architecting Working Group provides one such outlet. Still others will be needed for further growth.

In summary, all the indicators point to a future of high promise and value to all stakeholders.

References

1. *Webster's II New Riverside University Dictionary,* Riverside Publishing, Boston, MA, 1984, p. 939.
2. *The Institute of Electrical and Electronic Engineers (IEEE) Transactions on Systems, Man and Cybernetics,* Institute of Electrical and Electronic Engineers, New York; *IEEE Transactions on Aerospace and Electronic Systems,* Institute of Electrical and Electronic Engineers, New York; *J. Am. Inst. Aeronautics Astronautics,* American Institute of Aeronautics and Astronautics, Washington, D.C.
3. Machol, R.E., *Systems Engineering Handbook,* McGraw-Hill, New York, 1965; Chestnut, H., *System Engineering Methods,* John Wiley & Sons, New York, 1967; Blanchard, B.S. and Fabrycky, *Systems Engineering and Analysis,* Prentice-Hall, Englewood Cliffs, NJ, 1981.
4. *Proc. 1st Annu. Symp. National Council on Systems Engineering,* Seattle, WA, 1990; *J. Nat. Counc. Syst. Eng.,* Inaugural Issue, National Council on Systems Engineering, Seattle, WA, 1994; *Tools for System Engineering,* a brochure, Ascent Logic Corporation, San Jose, CA, 1995.
5. Rechtin, E., *Systems Architecting, Creating & Building Complex Systems,* Prentice-Hall, Englewood Cliffs, NJ, 1991, 299.
6. See the Inaugural Issue of *Systems Engineering, J. Nat. Counc. Syst. Eng.,* 1(1), July/September 1994.
7. See "Toward a Recommended Practice for Architectural Description" of the IEEE SESC Architecture Planning Group, April 9, 1996.

8. Rechtin, E., *Systems Architecting, Creating & Building Complex Systems,* Prentice-Hall, Englewood Cliffs, NJ, 1991.

9. *Doors,* A Brochure, Zycad Corporation, Freemont, CA, 1995; *SEDA — System Engineering and Design Automation,* a brochure, Nu Thena Systems, McLean, VA, 1995.

10. *EIA Standard IS-632,* Electronic Industries Association Publication, Washington, D.C., 1994.

11. Eisner, H., *Computer Aided Systems Engineering,* Prentice-Hall, Englewood Cliffs, NJ, 1988.

12. Perry, W.J., *Specifications and Standards — A New Way of Doing Business,* DODI 5000.2, Part 6, Section I, Department of Defense, Washington, D.C., 1995.

13. *IEEE P1220 Trail Use Standard for Application and Management of the Systems Engineering Process,* IEEE Standards Department, New York, 1994.

14. Martin, J.N., American national standard on systems engineering, *NCOSE INSIGHT, Issue 9,* International Council on Systems Engineering, Seattle, WA, September 1995.

15. Hammer, M. and Champy, J., *Reengineering the Corporation,* Harper, New York, 1994; Hamel, G. and Prahalad, C.K., *Competing for the Future,* Harvard Business School Press, Boston, MA, 1994.

16. Honour, E.C., TQM development of a systems engineering process, in *Proc. 3rd Annual Symp. of the National Council Systems Engineering,* National Council on Systems Engineering, Sunnyvale, CA, 1993.

17. Sheard, S.A. and Margolis, E.M., Team structures for systems engineering in an IPT environment, in *Proc. 5th Annu. Symp. of the National Council on Systems Engineering,* National Council on Systems Engineering, Sunnyvale, CA, 1995.

18. *Systems Engineering Handbook.* Hughes Aircraft Company, Culver City, CA, 1994.

19. Flood, R.L. and Carson, E.R., *Dealing with Complexity, An Introduction to the Theory and Application of Systems Science,* Plenum Press, New York, 1988.

20. Klir, G.J., *Architecture of Problem Solving,* Plenum Press, New York, 1988.

21. Rubinstein, M.F., *Patterns of Problem Solving,* Prentice-Hall, Englewood Cliffs, NJ, 1975.

22. Churchman, C.W., *The Design of Inquiring Systems,* Basic Books, New York, 1971.

23. Nadler, G., *The Planning and Design Approach,* John Wiley & Sons, New York, 1981.

24. Rechtin, E., *Systems Architecting, Creating & Building Complex Systems,* Prentice-Hall, Englewood Cliffs, NJ, 1991.

25. Lang, J., *Creating Architectural Theory, The Role of the Behavioral Sciences in Environmental Design,* Van Nostrand Reinhold, New York, 1987.

26. Rowe, P.G., *Design Thinking,* MIT Press, Cambridge, 1987.

27. Losk, Pieronek, Cureton, Geis, and Carpenter graduate reports are unpublished but available through the USC School of Engineering, Los Angeles, CA 0089-1450.

28. Alexander, C., *Notes on the Synthesis of Form,* Harvard University Press, Cambridge, MA, 1964. The first in a series of books on the subject.

29. Kostof, S., *The Architect, Chapters in the History of the Profession*, Oxford University Press, New York, 1977.
30. Genesereth, M.R. and Nilsson, N.J., *Logical Foundations of Artificial Intelligence*, Morgan Kaufmann Publishers, Los Altos, CA, 1987.
31. Newell, A., *Unified Theories of Cognition*, Harvard University Press, Cambridge, MA, 1990.
32. Simon, H.A., *The Sciences of the Artificial*, MIT Press, Cambridge, 1988.
33. Brooks, F.P., Jr., *The Mythical Man-Month*, Addison-Wesley, Reading, MA, 1982.
34. Garlan, D. and Shaw, M., *An Introduction to Software Architecture*, Carnegie Mellon University, Pittsburgh, PA, 1993.
35. Booton, R.C., Jr. and Ramo, S., The development of systems engineering, *IEEE Transac. Aerospace Electronic Syst.*, AES-20, 4, 306, July 1984.

Appendices

Appendix A

Heuristics for systems-level architecting

Experience is the hardest kind of teacher.

It gives you the test first and the lesson afterward.

(Susan Ruth 1993)

Introduction: organizing the list

The heuristics to follow were selected from Rechtin 1991, the *Collection of Student Heuristics in Systems Architecting*, 1988-1993,[1] and from subsequent studies in accordance with the selection criteria of Chapter 2. The list is intended as a tool store for top-level systems architecting. Heuristics continue to be developed and refined not only for this level, but for domain-specific applications as well, often migrating from domain-specific to system level and vice versa.*

For easy search and use, the heuristics are grouped by architectural task and categorized by being either descriptive or prescriptive; that is, by whether they describe an encountered situation or prescribe an architectural approach to it, respectively.

There are over 180 heuristics in the listing to follow, far too many to study at any one time. Nor were they intended to be. The listing is intended to be scanned as one would scan software tools on software store shelves, looking for ones that can be useful immediately but remembering that others are also there. Although some are variations of other heuristics, the vast majority stand on their own, related primarily to others in the near vicinity on the list. Odds are that the reader will find the most interesting heuristics in clusters, the location of which will depend on the reader's interests at the time. The section headings

---

* The manufacturing, social, communication, software, managment, business, and economics fields are particularly active in proposing and generating heuristics — though they usually are called principles, laws, rules, or axioms.

are by architecting task. A "D" signifies a descriptive heuristic; a "P"signifies a prescriptive one. When readily apparent, prescriptions are grouped by insetting under appropriate descriptions or alternate prescriptions; otherwise not. In the interests of brevity, an individual heuristic is listed in the task where it is most likely to be used most often. As noted in Chapter 2, some 20% can be tied to related ones in other tasks.

A major difference between an heuristic and an unsupported assertion is the credibility of the source. To the extent possible, the following heuristics are credited to the individuals who, to the authors' knowledge, first proposed and illustrated them. When not credited explicitly, they are as expressed by Rechtin and/or Maier. To further aid the reader in judging credibility or in finding the original context of the source, each heuristic is given a symbol indicating that the heuristic is from:

[ ]   *An informal discussion* with the individual indicated, unpublished.

( )   *A formal, dated source, with examples,* located in the USC MS-SAE program archive, especially in the *Collection of Student Heuristics in Systems Architecting, 1988-93.* For further information, contact the Master of Science Program in Systems Architecture and Engineering, USC School of Engineering, University Park, Los Angeles, CA 90089-1450.

*   Rechtin 1991, where it is sourced more formally. By permission of Prentice-Hall Inc., Englewood Cliffs, NJ.

Bold face Used for key words to aid in quick search.

Heuristic tool list

Multitask heuristics

D **Performance, cost, and schedule** cannot be specified independently. At least one of the three must depend on the others.*

D With few exceptions, schedule **delays** will be **accepted** grudgingly; cost **overruns** will **not,** and for good reason.

D The **time to completion** is proportional to the ratio of the time spent to the time planned to date. The greater the ratio, the longer the time to go.

---

* As indicated in the introduction to this appendix, an asterisk indicates that this heuristic is taken from Rechtin, E., *Systems Architecting, Creating & Building Complex Systems,* Prentice-Hall, Englewood Cliffs, NJ, 1991. With permission of Prentice-Hall, Englewood Cliffs, NJ 07632.

D **Relationships** among the elements are what give systems their added value.*

D **Efficiency** is inversely proportional to **universality**. (Douglas R. King, 1992)

D **Murphy's Law,** "If anything can go wrong, it will."*

   P **Simplify.** Simplify. Simplify.*

   P The first line of defense against complexity is simplicity of design.

   P Simplify, combine, and eliminate. (Suzaki, 1987)

   P Simplify with smarter elements. (N. P. Geiss, 1991)

   P The most **reliable part** on an airplane is the one that isn't there — because it isn't needed. [DC-9 Chief Engineer, 1989]

D One person's **architecture** is another person's **detail**. One person's system is another's component. [Robert Spinrad, 1989]*

P In order to **understand anything,** you must not try to understand everything. (Aristotle, 4th cent. B.C.)

P Don't confuse the functioning of the parts for the functioning of the system. (Jerry Olivieri, 1992)

D In general, each **system level** provides a context for the level(s) below. (G. G. Lendaris, 1986)

   P Leave the **specialties** to the specialist. The level of detail required by the architect is only to the depth of an element or component critical to the system as a whole. (Robert Spinrad, 1990) But the architect must have **access** to that level and know, or be informed, about its criticality and status. (Rechtin, 1990)

   P Complex systems will develop and **evolve** within an overall architecture much more rapidly if there are **stable intermediate** forms than if there are not. (Simon, 1969)*

D Particularly for social systems, it's the **perceptions**, not the facts, that count.

D In introducing technological and **social** change, **how** you do it is often more important than **what** you do.*

   P If social cooperation is required, the **way** in which a system is **implemented** and introduced must be an **integral part** of its architecture.*

D If the **politics** don't fly, the hardware never will. (Brenda Forman, 1990)

   D Politics, not technology, sets the limits of what technology is allowed to achieve.

   D Cost rules.

   D A strong, coherent constituency is essential.

   D Technical problems become political problems.

   D There is no such thing as a purely technical problem.

   D The best engineering solutions are not necessarily the best political solutions.

D **Good products** are not enough. Implementations matter. (Morris and Ferguson, 1993)

    P To remain competitive, determine and **control the keys** to the architecture from the very beginning.

Scoping and planning

> *The beginning is the most important part of the work.* (**Plato** 4th cent. B.C.)

> *Scope! Scope! Scope!* (William C. Burkett, 1992)

D **Success** is defined by the **beholder**, not by the architect.*

    P The most important single element of success is to **listen** closely to what the customer perceives as his requirements and to have the will and ability to be responsive. (J. E. Steiner, 1978)*

    P **Ask early** about how you will **evaluate** the success of your efforts. (Hayes-Roth et al., 1983)

    P For a system to meet its **acceptance criteria** to the satisfaction of all parties, it must be architected, designed, and built to do so — no more and no less.*

    P Define how an **acceptance** criterion is to be certified at the same time the criterion is established.*

    D Given a **successful** organization or system with valid criteria for success, there are some things **it cannot do** — or at least not do well. Don't force it!

    P The **strengths** of an organization or system in one context can be its **weaknesses** in another. Know when and where!*

    D There's nothing like being the **first success**.*

    P If at first you don't succeed, but the architecture is sound, try, try again. Success sometimes is where you find it. Sometimes it finds you.*

    D A system is successful when the **natural intersection** of technology, politics, and economics is found. (A. D. Wheelon, 1986)*

    D Four questions, **the Four Who's**, need to be answered as a self-consistent set if a system is to succeed economically; namely, who benefits?, who pays? who provides? and, as appropriate, who loses?

D **Risk** is (also) defined by the **beholder**, not the architect.

    P If being absolute is impossible in estimating system risks, then be relative.*

D **No** complex system can be **optimum** to all parties concerned, nor all functions optimized.*

P Look out for hidden agendas.*

P It is sometimes more important to know **who** the customer is than to know what the customer **wants**. (Whankuk Je, 1993)

D The phrase, "**I hate it**," is direction. (Lori I. Gradous, 1993)

P Sometimes, but not always, the best way to solve a difficult problem is to **expand** the problem itself.*

   P Moving to a **larger purpose** widens the range of solutions. (Gerald Nadler, 1990)

   P Sometimes it is necessary to **expand the** *concept* in order to simplify the problem. (Michael Forte, 1993)

   P [If in difficulty,] **Reformulate** the problem and re-allocate the system functions. (Norman P. Geis, 1991)

   P Use **open** architectures. You will need them once the market starts to respond.

P Plan to **throw one away**. You will anyway. (F. P. Brooks, Jr., 1982)

   P You can't avoid **redesign**. It's a natural part of design.*

P Don't make an architecture **too smart** for its own good.*

D Amid a **wash of paper**, a small number of documents become critical pivots around which every project's management revolves. (F. P. Brooks, Jr., 1982)*

   P Just because it's written, doesn't make it so. (Susan Ruth, 1993)

D In architecting a new [software] program all the **serious mistakes** are made in the **first day**. [Spinrad, 1988]

   P The most **dangerous** assumptions are the **unstated** ones. (Douglas R. King, 1991)

   D Some of the **worst** failures are **systems** failures.

D In architecting a new [aerospace] system, by the time of the first **design review**, performance, cost, and schedule have been **predetermined**. One might not know what they are yet, but to first order all the critical assumptions and choices have been made which will determine those key parameters.*

P **Don't assume** that the original statement of the problem is necessarily the best, or even the right, one.*

   P **Extreme** requirements, expectations, and predictions should remain under challenge throughout system design, implementation, and operation.

   P Any extreme requirement must be intrinsic to the system's design philosophy and must validate its selection. "Everything must pay its way on to the airplane." [Harry Hillaker 1993]

   P **Don't assume** that previous **studies** are necessarily complete, current, or even correct. (James Kaplan, 1992)

   P Challenge the process and solution, for surely someone else will do so. (Kenneth L. Cureton, 1991)

   P Just because it worked in the past there's no guarantee that it will work now or in the future. (Kenneth L. Cureton, 1991)

P Explore the situation from more than one point of view. A seemingly impossible situation might suddenly become transparently simple. (Christopher Abts, 1988)

P **Work forwards and backwards**. (A set of heuristics from **Rubinstein** 1975)*
    Generalize or specialize.
    Explore multiple directions based on partial evidence.
    Form stable substructures.
    Use analogies and metaphors.
    Follow your emotions.

P Try to hit a solution that, at worst, won't put you **out of business**. (Bill Butterworth as reported by Laura Noel, 1991)

P The **order** in which decisions are made can change the architecture as much as the decisions themselves. (Rechtin, 1975, IEEE SPECTRUM)

P Build in and maintain **options** as long as possible in the design and build of complex systems. You will need them. OR ... Hang on to the agony of decision as long as possible. [Robert Spinrad, 1988]*
    P Successful architectures are **proprietary, but open**. [Morrison and Ferguson, 1993]

D Once the architecture begins to take shape, the sooner contextual constraints and **sanity checks** are made on assumptions and requirements, the better.*

D Concept **formulation** is complete when the **builder** thinks the system can be built to the **client's** satisfaction.*

D The **realities** at the end of the conceptual phase are not the models but the **acceptance criteria**.*

P Do the **hard parts** first.
    P Firm **commitments** are best made after the **prototype works**.

Modeling (see also Chapters 3 and 4)

P If you can't analyze it, don't build it.

D Modeling is a **craft** and at times an art. (William C. Burkett, 1994)

D A **vision** is an **imaginary architecture** ... no better, no worse than the rest of the models. (M. B. Renton, Spring 1995)

D From psychology: if the concepts in the mind of one person are very different from those in the mind of the other, there is no **common model** of the topic and no communication. [Taylor, 1975] OR ... From telecommunications: The **best receiver** is one that contains an internal model of the transmitter and the channel. [Robert Parks and Frank Lehan, 1954]*

D A model is not **reality**.*
    D The **map** is not the territory. (Douglas R. King, 1991)*

    P Build **reality checks** into model-driven development. [Larry Dumas, 1989]*

    P Don't believe **nth order consequences** of a first order [cost] model. [R. W. Jensen, *circa* 1989]

  D Constants aren't and variables don't. (William C. Burkett, 1992)

  D One insight is worth a thousand **analyses**. (Charles W. Sooter, 1993)

    P Any war game, systems analysis, or study whose results can't easily be explained on the back of an envelope is not just worthless, it is probably dangerous. [Brookner-Fowler, *circa* 1988]

  D Users develop **mental models** of systems based [primarily] upon the user-to-system interface. (Jeffrey H. Schmidt)

  D If you can't explain it in **five minutes**, either you don't understand it or it doesn't work. (Darcy McGinn, 1992, from David Jones)

  P The **eye** is a fine **architect**. Believe it. [Wernher von Braun, 1950]

  D A good solution somehow **looks nice**. (Robert Spinrad, 1991)

    P **Taste**: an aesthetic feeling which will accept a solution as right only when no more direct or simple approach can be envisaged. [Robert Spinrad, 1994]

    P Regarding **intuition**, trust but **verify**. (Jonathan Losk, 1989)

Prioritizing (trades, options, and choices)

  D In any resource-limited situation, the **true value** of a given service or product is determined by what one is willling to **give up** to obtain it.

  P When choices must be made with unavoidably inadequate information, **choose** the best available and then **watch** to see whether future solutions appear faster than future problems. If so, the choice was at least adequate. If not, go back and **choose** again.*

  P When a decision makes sense through several different **frames**, it's probably a good decision. (J. E. Russo, 1989)

  D The **choice** between architectures may well depend upon which set of **drawbacks** the client can handle best.*

  P If trade results are inconclusive, then the wrong selection criteria were used. Find out [again] what the customer wants and why they want it, then repeat the trade using those factors as the [new] selection criteria. (Kenneth Cureton, 1991)

  P The triage: let the dying die. Ignore those who will recover on their own. And treat only those who would die without help.*

  P Every once in a while you have to go back and see what the real world is telling you. [Harry Hillaker, 1993]

Aggregating ("chunking")

P **Group** elements that are strongly **related** to each other, **separate** elements that are unrelated.

D Many of the **requirements** can be **brought together** to complement each other in the total design solution. Obviously the more the design is put together in this manner, the more probable the overall success. (J. E. Steiner, 1978)

P Subsystem interfaces should be drawn so that each subsystem can be implemented independently of the specific implementation of the subsystems to which it interfaces. (Mark Maier, 1988)

P Choose a configuration with **minimal communications** between the subsystems.

  P Choose the elements so that they are as independent as possible; that is, elements with low external complexity (low coupling) and high internal complexity (high cohesion). (Christopher Alexander, 1964, modified by Jeff Gold, 1991)*

  P Choose a configuration in which local activity is high speed and global activity is slow change. (Courtois 1985)*

P Poor aggregation results in **gray** boundaries and **red** performance. (M. B. Renton, Spring 1995)

  P **Never aggregate** systems that have a **conflict** of interest; partition them to ensure checks and balances. (Aubrey Bout, 1993)

  P **Aggregate** around "**testable**" subunits of the product; **partition** around logical **subassemblies**. (Ray Cavola, 1993)

  P Iterate the partition/aggregation procedure until a model consisting of **7 ± 2 chunks** emerge. (Moshe F. Rubinstein, 1975)

  P The **optimum number** of architectural elements is the amount that leads to distinct **action**, not general planning. (M. B. Renton, Spring 1995)

P System structure should resemble functional structure.*

  P Except for good and sufficient reasons, **functional and physical** structuring should match.*

  P The architecture of a **support** element must **fit** that of the system which it supports. It is easier to match a support system to the human it supports than the reverse.*

P **Unbounded limits** on element behavior may be a **trap** in unexpected scenarios. [Bernard Kuchta, 1989]*

Partitioning (decompositioning)

P Do not **slice** through regions where **high rates** of information exchange are required. (computer design)*

D The greatest **leverage** in architecting is at the **interfaces**.*

P **Guidelines** for a good quality interface specification: they must be simple, unambiguous, complete, concise, and focus on substance. Working documents should be the same as customer deliverables; that is, always use the customer's language, not engineering jargon. [Harry Hillaker, 1993]

P The **efficient architect**, using contextual sense, continually looks for likely **misfits** and redesigns the architecture so as to eliminate or minimize them. (Christopher Alexander, 1964)* It is inadequate to architect up to the boundaries or **interfaces** of a system; one must architect across them. (Robert Spinrad, as reported by Susan Ruth, 1993)

P Since boundaries are inherently limiting, look for solutions outside the boundaries. (Wolf, Steven, 1992)

P Be prepared for **reality** to add a few interfaces of its own.*

P Design the structure with good **"bones"**.*

P Organize personnel tasks to **minimize** the **time** individuals spend interfacing. (Tausworthe, 1988)*

Integrating

D **Relationships** among the elements are what give systems their **added value**.*

P The greatest leverage in system architecting is at the interfaces.*

P The greatest **dangers** are also at the interfaces. [Raymond, 1988]

P Be sure to ask the question, "What is the worst thing that other elements could do to you across the interface?" [Kuchta, 1989]

D Just as a piece and its template must match, so must a system and the resources which make, test, and operate it. Or, more briefly, the **product and process** must match. Or, by extension, a system architecture cannot be considered complete lacking a suitable match with the process architecture.*

P When confronted with a particularly difficult interface, try changing its **characterization**.*

P Contain **excess energy** as close to the source as possible.*

P Place barriers in the paths between energy sources and the elements the energy can damage. (Kjos, 1988)*

Certifying (system integrity, quality, and vision)

D As **time to delivery** decreases, the **threat** to functionality increases. (Steven Wolf, 1992)

P If it is a good design, **insure** that it stays **sold**. (Dianna Sammons, 1991)

D Regardless of what has gone before, the **acceptance criteria** determine what is actually built.*

    D The number of **defects remaining** in a (software) system after a given level of test or review (design review, unit test, system test, etc.) is proportional to the **number found** during that test or review.

    P **Tally** the defects, analyze them, **trace** them to the source, make corrections, keep a **record** of what happens afterwards, and keep **repeating** it. (Deming)

    P **Discipline.** Discipline. Discipline. (Douglas R. King, 1991)

    P The principles of **minimum communications** and proper partitioning are key to system testability and **fault isolation**. (Daniel Ley, 1991)*

    P **The five why's** of Toyota's lean manufacturing. (To find the basic cause of a defect, keep asking "why" from effect to cause to cause five times.)

D The test **setup** for a system is itself a system.*

    P The test system should always allow a part to pass or fail on its own merit. [James Liston, 1991]*

    P To be tested, a system must be designed to be tested.*

D An element "**good enough**" in a small system is unlikely to be good enough in a more complex one.*

D Within the same class of products and processes, the **failure rate** of a product is linearly proportional to its **cost**.*

D The **cost** to find and **fix** an inadequate or failed part increases by an **order of magnitude** as it is successively incorporated into higher levels in the system.

    P The least expensive and most effective place to find and fix a problem is at its source.

D Knowing a **failure has occurred** is more important than the actual failure. (Kjos, 1988)

D **Mistakes** are **understandable**, failing to report them is **inexcusable**.

D Recovery from failure or flaw is not complete until a specific mechanism, **and no other**, has been shown to be the cause.*

D Reducing **failure rate** by each **factor of two** takes as much effort as the original development.*

D **Quality** can't be tested in, it has to be **built in**.*

    D You can't achieve quality ... unless you specify it. (Deutsch, 1988)

    P Verify the quality close to the source. (Jim Burruss, 1993)

    P The five why's of Japan's lean manufacturing. (Hayes et al., 1988)

    D High **quality**, reliable systems are produced by high quality architecting, engineering, design and manufacture, **not by inspection**, test, and rework.*

    P Everyone in the development and production line is both a customer and a supplier.

    D Next to interfaces, the greatest **leverage** in architecting is in aiding the recovery from, or exploitation of, **deviations** in system performance, cost, or schedule.*

Assessing performance, cost, schedule, and risk

    D A good design has benefits in more than one area. (Trudy Benjamin, 1993)

    D System quality is defined in terms of customer satisfaction, not requirements satisfaction. (Jeffrey Schmidt, 1993)

    D If you think your **design** is perfect, it's only because you haven't shown it to **someone else**. [Harry Hillaker, 1993]

        P Before proceeding too far, **pause and reflect!** Cool off periodically and seek an independent review. (Douglas R. King, 1991)

    D Qualification and **acceptance** tests must be both definitive and **passable**.*

        P High **confidence**, not test completion, is the **goal** of successful qualification. (Daniel Gaudet, 1991)

        P Before ordering a **test** decide what you will do if it is (1) **positive** or if (2) it is negative. If both answers are the same, **don't do** the test. (R. Matz, M.D., 1977)

    D "**Proven**" and "**state of the art**" are mutually **exclusive** qualities. (Lori I. Gradous, 1993)

    D The **bitterness** of **poor performance** remains long after the sweetness of low prices and prompt delivery are forgotten. (Jerry Lim, 1994)

    D The **reverse of diagnostic** techniques are good architectures. (M.B. Renton, 1995)

    D Unless everyone who **needs to know** does know, somebody, somewhere will foul up.

        P Because there's no such thing as immaculate communication, don't ever stop **talking** about the system. (Losk, 1989)*

    D Before it's tried, it's **opinion**. After it's tried, it's **obvious**. (Wm. C. Burkett, 1992)

    D Before the **war** it's opinion. After the war, it's too late! (Anthony Cerveny, 1991)

    D The first **quick look** analyses are often **wrong**.*

    D In correcting system deviations and failures it is important that all the participants know not only **what** happened and how it happened but **why** as well.*

        P Failure reporting without a **close out** system is meaningless. (April Gillam, 1989)

P Common, if undesirable, responses to **indeterminate out-comes** or failures:*
if it **ain't broke**, don't fix it.
Let's **wait and see** if it goes away or happens again.
It was just a **random** failure. One of those things.
Just treat the **symptom**. Worry about the cause later.
Fix everything that might have caused the problem.
Your **guess** is as good as mine.

D Chances for recovery from a **single failure** or flaw, even with complex consequences, are fairly good. Recovery from **two or more** independent failures is unlikely in real time and uncertain in any case.*

Rearchitecting (evolving, modifying, and adapting)

> *The test of a good architecture is that it will last.*
> *The sound architecture is an enduring pattern.*

[Robert Spinrad, 1988]

P The team that created and built a presently successful product is often the best one for its evolution — but seldom for creating its replacement.

D If you **don't understand** the existing system, you can't be sure you're rearchitecting a **better** one. (Susan Ruth, 1993)

P When implementing a change, keep some elements constant to provide an **anchor** point for people to cling to. (Jeffrey H. Schmidt, 1993)

  P In large, mature systems, **evolution** should be a process of **ingress** and **egress**. (IEEE 1992, Jeffrey Schmidt, 1992)

  P Before the change, it is your opinion. After the change it is your problem. (Jeffrey Schmidt, 1992)

D Unless constrained, **rearchitecting** has a natural tendency to proceed unchecked until it results in a substantial transformation of the system. (Charles W. Sooter, 1993)

D Given a change, if the anticipated actions don't occur, then there is probably an invisible barrier to be identified and overcome. (Susan Ruth, 1993)

Exercises

Exercise: What favorite heuristics, rules of thumb, facts of life, or just plain common sense do you apply to your own day-to-day living — at work, at home, at play? What heuristics etc.

have you heard on TV or the radio; for example, on talk radio, action TV, children's programs? Which ones would you trust?

Exercise: Choose a system, product, or process with which you are familiar and assess it using the appropriate foregoing heuristics. What was the result? Which heuristics are or were particularly applicable? What further heuristics were suggested by the system chosen? Were any of the heuristics clearly incorrect for this system? If so, why?

Exercise: Try to spot heuristics and insights in the technical literature. Some are easy; they are often listed as principles or rules. The more difficult ones are buried in the text but contain the essence of the article or state something of far broader application than the subject of the piece.

Exercise: Try to create an heuristic of your own — a guide to action, decision making, or to instruction of others.

Notes and References

1. Rechtin, E., Ed., University of Southern California, Los Angeles, March 15, 1994 (unpublished but available to students and researchers on request).
2. Hayes, R.H., Wheelwright, S.C., and Clark, K.B., *Dynamic Manufacturing, Creating the Learning Organization*, Free Press, New York, 1988.

Appendix B

Reference list for systems architecting

The following list is offered as a brief guide to books that would be particularly appropriate to an architecting library.

Architecting background

Alexander, C., *A Pattern Language: Towns, Buildings, Construction*, Oxford University Press, New York, 1977.

Alexander, C., *The Timeless Way of Building*, Oxford University Press, New York, 1979.

Alexander, C., *Notes on the Synthesis of Form*, Harvard University Press, Cambridge, MA, 1964.

Kostof, Spiro, *The Architect*, Oxford University Press, New York, 1977.

Lang, Jon, *Creating Architectural Theory*, van Nostrand Reinhold Company, 1987.

Rowe, P.G., *Design Theory*, MIT Press, Cambridge, MA, 1987.

Vitruvius, *The Ten Books on Architecture*, translated by Morris Hicky Morgan, Dover Publications, Mineola, NY, 1960.

Management

Augustine, N.R., *Augustine's Laws*, AIAA, Inc., New York, 1982.

Deal, Terrence E. and Kennedy, A.A., *Corporate Cultures, The Rites and Rituals of Corporate Life*, Addison-Wesley, Reading, MA, 1988.

DeMarco, T. and Lister, T., *Peopleware: Productive Projects and Teams*, Dorset House, New York, 1987.

Juran, J.M., *Juran on Planning for Quality*, The Free Press, A Division of Macmillan, New York, 1988.

Modeling

Eisner, H., *Computer Aided Systems Engineering*, Prentice-Hall, Englewood Cliffs, NJ, 1988.

Hatley, D.J., and Pirbhai, I., *Strategies for Real-Time System Specification*, Dorset House, New York, 1988.

J. Rumbaugh, M. Blaha, W. Premerlani, F. Eddy, and W. Lorensen, *Object-Oriented Modeling and Design*, Prentice-Hall, Englewood Cliffs, NJ, 1988.

Ward, P.T. and Mellor, S.J, *Structured Development for Real-Time Systems, Volume 1: Introduction and Tools*, Yourdon Press (Prentice-Hall), Englewood Cliffs, NJ, 1985.

Specialty areas

Baudin, M., *Manufacturing Systems Analysis*, Yourdon Press Computing Series (Prentice-Hall), Englewood Cliffs, NJ, 1990.

Hayes, Robert H., Wheelwright, S.C., and Clark, K.B, *Dynamic Manufacturing*, The Free Press, A Division of Macmillan, New York, 1988.

Miller, J.G., *Living Systems*, McGraw Hill, New York, 1978.

Simon, H.A., *Sciences of the Artificial*, MIT Press, Cambridge, 1981.

Thome, B., editor, *Systems Engineering: Principles and Practice of Computer-Based Systems Engineering*, John Wiley, Baffins Lane, Chichester, Wiley Series on Software Based Systems, 1993.

Software

Boehm, B., *Software Engineering Economics*, Prentice-Hall, Englewood Cliffs, NJ, 1981.

Brooks, F.P., Jr., *The Mythical Man-month, Essays on Software Engineering, 20th Anniversary Edition*, Addison-Wesley, Reading, MA, 1995.

Gajski, D.D., V.M. Milutinović, H.J. Siegel and B.P. Furht, *Computer Architecture* (Tutorial), The Computer Society of the IEEE, 1987.

Gamma, E., et al, *Design Patterns: Elements of Reusable Object-Oriented Software Architecture*, Addison Wesley, Reading, MA, 1994.

Deutsch, M.S. and Willis, R.R., *Software Quality Engineering*, Prentice-Hall, Englewood Cliffs, NJ, 1988.

Software Productivity Consortium, ADARTS Guidebook, SPC-94040-CMC, Version 2.00.13, Vols. 1-2, Herndon, VA, September, 1991.

Shaw, M. and Garlan, D., *Software Architecture: Perspectives on an Emerging Discipline*, Prentice-Hall, Englewood Cliffs, NJ, 1996.

Yourdon, E. and Constantine, L.L., *Structured Design: Fundamentals of a Discipline of Computer Program and Systems Design*, Yourdon Press (Prentice-Hall), Englewood Cliffs, NJ, 1979.

System sciences

Flood, R.L. and Carson, E.R., *Dealing with Complexity, an Introduction to the Theory and Application of System Sciences*, Plenum Press, New York, 1988.

Genesereth, M.R. and Nilsson, N.J., *Logical Foundations of Artificial Intelligence*, Morgan Kaufmann, San Francisco, CA, 1987.

Gerstein, Dean R., et al. Editors, *The Behavioral and Social Sciences, Achievements and Opportunities*, National Academy Press, Washington, DC, 1988.

Klir, G.J., *Architecture of Systems Problem Solving*, Plenum Press, New York, 1985.

System thinking

Arbib, M.A., *Brains, Machines, and Mathematics,* 2nd ed., Springer-Verlag, New York, 1987.

Beam, Walter R., *Systems Engineering, Architecture and Design,* McGraw-Hill, New York, 1990.

Boorstin, Daniel J., *The Discoverers,* Vintage Books, New York, 1985.

Boyes, J.L., Editor, *Principles of Command and Control,* AFCEA International Press, Washington, DC, 1987.

Davis, S.M., *Future Perfect,* Addison Wesley, Reading, MA, 1987.

Gause, Donald C. and Weinberg, G. M., *Exploring Requirements, Quality Before Design,* Dorset House, New York, 1989.

Hofstadter, D.R., *Gödel, Escher, Bach: an eternal golden braid,* Vintage Books, New York, 1980.

Norman, Donald A., *The Psychology of Everyday Things,* Basic Books, New York, 1988.

Pearl, Judea, *Heuristics,* Reading, MA, 1984.

Rechtin, E., *Systems Architecting: Creating and Building Complex Systems,* Prentice-Hall, Englewood Cliffs, NJ, 1991.

Rubinstein, Moshe F., *Patterns of Problem Solving,* Prentice-Hall, Englewood Cliffs, NJ, 1975.

Weinberg, Gerald M., *Rethinking Systems Analysis and Design,* Dorset House, New York, 1988.

Glossary

Glossary

The fields of system engineering and systems architecting are sufficiently new that many terms have not yet been standardized. Common usage is often different among different groups and in different contexts. However, *for the purposes of this book*, we have provided the meanings of the following terms:

Abstraction   A representation in terms of presumed essentials, with a corresponding suppression of the non-essential.

ADARTS   Ada-based Design Approach for Real Time Systems. A software development method (including models, processes, and heuristics) developed and promoted by the Software Productivity Consortium.

Aggregation   The gathering together of closely related elements, purposes, or functions.

Architecting   The process of creating and building architectures. Depending on one's perspective, architecting may or may not be seen as a separable part of engineering. Those aspects of system development most concerned with conceptualization, objective definition, and certification for use.

Architecture   The structure — in terms of components, connections, and constraints — of a product, process, or element.

Architecture, open   An architecture designed to facilitate addition, extension, or adaptation for use.

Architecture (communication, software, hardware, etc.)   The architecture of the particular designated aspect of a large system.

Architectural style   A form or pattern of design with a shared vocabulary of design idioms and rules for using them.[1]

ARPANET/INTERNET   The global computer internetwork, principally based on the TCP/IP packet communications protocol. The ARPANET was the original prototype of the current INTERNET.

Certification   A formal, but not necessarily mathematical, statement that defined system properties or requirements have been met.

Client   The individual or organization that pays the bills. May or may not be the user.

Complexity   A measure of the numbers and types of inter-relationships among system elements. Generally speaking, the more complex a system, the more difficult it is to design, build, and use.

Deductive reasoning   Proceeding from an established principle to its application.

Design   The detailed formulation of the plans or instructions for making a defined system element; a follow-on step to systems architecting and engineering.

Domain   A recognized field of activity and expertise, or of specialized theory and application.

Engineering   Creating cost-effective solutions to practical problems by applying scientific knowledge to building things in the service of mankind.[1] May or may not include the art of architecting.

Engineering, concurrent   Narrowly defined (here) as the process by which product designers and manufacturing process engineers work together to create a manufacturable product.

Heuristic   A guideline for architecting, engineering, or design. Lessons learned expressed as a guideline. A natural language abstraction of experience which passes the tests of Chapter 2.

Heuristic, descriptive   A heuristic that describes a situation.

Heuristic, prescriptive   A heuristic that prescribes a course of action.

IEEE P 1220   An Institute of Electrical and Electronic Engineers standard for system engineering.

Inductive reasoning   Extrapolating the results of examples to a more general principle.

Manufacturing, flexible   Creating different products on demand using the same manufacturing line. In practice, all products on that line come from the same family.

Manufacturing, lean   An efficient and cost-effective manufacturing or production system based on ultraquality and feedback.[2]

MBTI   Meyer-Briggs Type Indicator. A psychological test for indicating the temperaments associated with selected classes of problem solving.[3]

Metaphor   A description of an object or system using the terminology and properties of another, for example, the desktop metaphor for computerized document processing.

MIL-STD   Standards for defense system acquisition and development.

Model   An abstracted representation of some aspect of a system.

Model, satisfaction   A model which predicts the performance of a system in language relevant to the client.

Modeling   Creating and using abstracted representations of actual systems, devices, attributes, processes, or software.

Models, integrated   A set of models, representing different views, forming a consistent representation of the whole system.

Normative method   A design or architectural method based on "what should be," that is, on a predetermined definition of success.

OMT <u>O</u>bject <u>M</u>odeling <u>T</u>echnique.   An object-oriented software development method.[4]

Objectives   Client needs and goals, however stated.

Paradigm   A scheme of things, a defining set of principles, a way of looking at an activity, for example, classical architecting.

Participative method   A design method based on wide participation of interested parties. Designing through a group process.

Partitioning   The dividing up of a system into subsystems.

Progressive design   The concept of a continuing succession of design activities throughout product or process development. The succession progressively reduces the abstraction of the system through models until physical implementation is reached and the system used.

Purpose   A reason for building a system.

Rational method   A design method based on deduction from the principles of mathematics and science.

Requirement   An objective regarded by the client as an absolute; that is, either passed or not.

Scoping   Sizing; defining the boundaries and/or defining the constraints of a process, product, or project.

Spiral   A model of system development which repeatedly cycles from function to form, build, test, and back to function. Originally proposed as a risk-driven process, particularly applicable to software development with multiple release cycles.

System   A collection of things or elements which, working together, produce a result not achievable by the things alone.

Systems, builder-architected   Systems architected by their builders, generally without a committed client.

Systems, feedback   Systems which are strongly affected by feedback of the output to the input.

Systems, form first   Systems which begin development with a defined form (or architecture) instead of a defined purpose. Typical of builder-architected systems.

Systems, politico-technical   Technological systems the development and use of which is strongly influenced by the political processes of government.

Systems, socio-technical   Technological systems the development and use of which is strongly affected by diverse social groups. Systems in which social considerations equal or exceed technical ones.

Systems engineering   A multi-disciplinary engineering discipline in which decisions and designs are based on their effect on the system as a whole.

Systems architecting   The art and science of creating and building complex systems. That part of systems development most concerned with scoping, structuring, and certification.

Systems architecting, the art of   That part of systems architecting based on qualitative heuristic principles and techniques, that is, on lessons learned, value judgments, and unmeasurables.

Systems architecting, the science of   That part of systems architecting based on quantitative analytic techniques, that is, on mathematics and science and measureables.

Technical decisions   Architectural decisions based on engineering feasibility.

Ultraquality   Quality so high that measuring it in time and at reasonable cost is impractical.[5]

Waterfall   A development model based on a single sequence of steps; typically applied to the making of major hardware elements.

Value judgments   Conclusions based on worth (to the client and other stakeholders).

View   A perspective on a system describing some related set of attributes. A view is represented by one or more models.

Zero defects   A production technique based on an objective of making everything perfectly. Related to the "everyone a supplier, everyone a customer" technique for eliminating defects at the source. Contrasts with acceptable quality limits in which defects are accepted providing they do not exceed specified limits in number or performance.

References

1. Shaw, M. and Garlan, D., *Software Architecture: Perspectives on an Emerging Discipline,* Prentice-Hall, Englewood Cliffs, NJ, p. 6, 19, 1996.
2. Womack, James P., Jones, Daniel T., and Roos, Daniel, *The Machine That Changed the World, The Story of Lean Production,* Harper Collins Publishers, New York, 1990.
3. Myers, Isabel, Briggs, Katheryn, and McCaulley, *Manual: A Guide to the Development and Use of the Myers-Briggs Type Indicator,* Palo Alto, CA: Consulting Psychologists Press, Palo Alto, CA, 1989.
4. Rumbaugh, J., et. al., *Object-Oriented Modeling and Design,* Prentice-Hall, Englewood Cliffs, NJ, 1991.
5. Rechtin, E., *Systems Architecting, Creating & Building Complex Systems,* Prentice-Hall, Englewood Cliffs, NJ, 1991, Ch. 8.

Indexes

Index of authors

Index by subject